David Hayes Agnew

Practical Anatomy

A new arrangement of the London dissector: with numerous modification and

additions, containing a concise description of the muscles, bloodvessels, nerves,

viscera, and ligaments of the human body as they appear on dissection

David Hayes Agnew

Practical Anatomy
A new arrangement of the London dissector: with numerous modification and additions, containing a concise description of the muscles, bloodvessels, nerves, viscera, and ligaments of the human body as they appear on dissection

ISBN/EAN: 9783337370312

Printed in Europe, USA, Canada, Australia, Japan

Cover: Foto ©berggeist007 / pixelio.de

More available books at **www.hansebooks.com**

PRACTICAL ANATOMY.

A NEW ARRANGEMENT

OF THE

LONDON DISSECTOR.

WITH

NUMEROUS MODIFICATIONS AND ADDITIONS,

CONTAINING

A CONCISE DESCRIPTION OF THE MUSCLES, BLOODVESSELS, NERVES, VISCERA, AND LIGAMENTS OF THE HUMAN BODY AS THEY APPEAR ON DISSECTION.

With Illustrations.

BY

D. HAYES AGNEW, M. D.,

Demonstrator of Anatomy and Assistant Lecturer on Clinical Surgery in the University of Pennsylvania, Surgeon to the Pennsylvania Hospital and the Will's Ophthalmic Hospital, etc. etc.

SECOND EDITION, REVISED.

PHILADELPHIA:

J. B. LIPPINCOTT & CO.

1868.

PREFACE

THE AMERICAN EDITOR.

THE Editor of the following pages, believing that a dissector suited to the wants of the American Student should be much more condensed than those in general use, has made the following arrangement of the London Dissector. In its preparation the arrangement has been entirely altered, and the nomenclature in many cases changed. The Ligamentous system: illustrations and numerous other additions in various parts of the work have been made; everything considered unnecessary erased; and the whole presented as near as possible in topographical order. The minute anatomy of parts, and the smaller ramifications of the nervous and vascular systems have been omitted, belonging as they do more properly to systematic treatises on Special Anatomy.

The work has been prepared with an eye single to the faithful economy of the Student's time.

D. HAYES AGNEW.

August, 1856.

(iii)

PREFACE

TO THE

SECOND EDITION

BY THE AMERICAN EDITOR.

THE first edition of this book on Practical Anatomy having been exhausted, and the favorable manner in which it has been received, have induced the preparation of a second edition.

The same care in this as in the former one has been observed, to compress the matter into as small a compass as would consist with perspicuity.

D. HAYES AGNEW,

1611 Chestnut Street.

August 22, 1867.

(v)

CONTENTS.

(vii)

viii CONTENTS.

CHAPTER X.

PRACTICAL ANATOMY.

CHAPTER I.

GENERAL OBSERVATIONS IN REFERENCE TO DISSECTING.

THE student about commencing the duties of the dissecting room, should be supplied with a good apron to protect his clothes from being soiled and a dissecting case of good quality. The knife should be held like a pen, and moved by the fingers and thumb, not by the wrist and elbow, and sometimes, as I have seen, by the shoulder. The student should practice its use also with the left hand; such an education will be of great advantage, especially in surgical operations. In the dissection of a part, it is a good rule not to turn off more of the skin than is necessary for the exposure of the part, and always to replace it again during the interval of dissection. Generally I think it better to turn off the structures in the order of superposition; as the skin, superficial fascia, deep fascia, etc. In dissecting a muscle the parts should be made tense, which is done by blocks properly placed under the subject. The cellular membrane which immediately invests the muscle, must be kept extended by the hand or forceps, and the edge of the knife carried steadily in the *direction of its fibres.* This rule admits of no exceptions, and should never be lost sight of by the student.

There are many operations which may be done on the subject without injuring the parts for dissection; such as the *ligation of arteries; passage* of instruments

2 (9)

through the *nasal fossæ* and their introduction into the *Eustachian tube; introduction of catheter*, both in the male and female, etc.; and these should be practiced with care. The grand object of the student in the examination of the human body, should be not only to recognize its constituents, but above all the relation which they sustain to each other; this is the only knowledge worth obtaining, and the want of it accounts for the blunders, confusion, and ignorance which are so frequently seen in the profession. Another point of much importance is *order and cleanliness*. Hence all drippings from the subject should be wiped up with the sponge, and no scraps allowed to accumulate about the table. Much apprehension is sometimes expressed by students about *poisoned wounds*. It is well if there be any abrasions of the skin to cover it with collodion before engaging in dissecting; but since the introduction of the chloride of zinc as an antiseptic all danger from this source is removed. Where an accidental cut is inflicted, the part should be washed, and well drawn by the mouth. The common practice of cauterizing with the nitras argenti, I think is in many cases positively injurious, unless well drawn first.

CHAPTER II.

DISSECTION OF THE HEAD.

Of the External Parts of the Head.

THE integuments of the head are thick, and covered with hair; under the cutis there is a cellular substance, which is much condensed, and closely connected with the epicranium, or expanded tendon of the occipito-frontalis. This connection renders the dissection of that muscle difficult.

An incision should be made from the root of the nose over the median line of the head, terminating behind at the occipital protuberance; a second from the commencement of the first, over the eyebrows to the external angular process of the frontal bone; a third from the occipital protuberance to the mastoid process of the temporal bone; by dissecting very carefully the flap outward, the entire muscle and intervening tendon or aponeurosis will be exposed.

The OCCIPITO-FRONTALIS is the only muscle which properly belongs to the hairy scalp; it is a single broad digastric muscle.

Arising, on each side of the head, fleshy and tendinous, from the transverse ridge of the occipital bone, as far forward as the mastoid process; it forms a broad, thin tendon, which covers the whole upper part of the cranium.

Inserted, fleshy, on each side, into the orbicularis palpebrarum, skin of the eyebrows, and the internal angular process of the os frontis and os nasi. The last insertion answers to the Pyramidalis Nasi of some books.

Situation. The tendon adheres firmly by cellular membrane to the skin, but very loosely to the pericranium or periosteum of the cranium, by a loose cellular layer, which might be called the *subaponeurotic fascia.* At its insertion it intermixes with the muscles of the upper part of the face.

Use. To pull the skin of the head backward, raise the eyebrows, and corrugate the skin of the forehead.

BLOODVESSELS seen in the dissection of the scalp are the FRONTAL, emerging from the internal angle of the orbit and passing up the middle of the head. *Supraorbital* out of the SUPRA-ORBITAL foramen; both branches of the OPHTHALMIC. On the side of the head the TEMPORAL, from the external carotid, behind the POSTERIOR OCCIPITAL, and POSTERIOR AURICULAR from the external carotid. All anastomose in the scalp.

The SUPRA-ORBITAL NERVES coming out of the supraorbital foramen. The FRONTAL NERVE, a branch of the ophthalmic, which last is the first branch of the fifth

pair; on the side, branches of the PORTIO DURA and
INFERIOR MAXILLARY NERVES; and behind, the OCCI-
PITALIS MINOR and MAJOR—the former from the superficial plexus of the neck, and the latter the main stalk
of the second cervical spinal nerve.

The muscles of the ear are of three classes.

1. The common muscles move the external ear; they
are not always so distinct as to admit of a clear demonstration.

(1) ATTOLLENS AUREM—*Arises* from the tendon of
the occipito-frontalis, and from the aponeurosis of the
temporal muscle.

Inserted into the upper part of the root of the cartilage of the ear, opposite to the antihelix.

Use. To draw the ear upward.

(2) ANTERIOR AURIS—*Arises*, thin and membranous,
from the posterior part of the zygomatic process of the
temporal bone.

Inserted into a small eminence on the back of the
helix, opposite to the concha.

Use. To draw the eminence a little forward and upward.

(3) The RETRAHENTES AURIS—*Arise*, by two or
three distinct slips, from the external and posterior part
of the mastoid process, immediately above the insertion
of the sterno-cleido mastoideus.

Inserted into that back part of the ear which is opposite to the septum, dividing the scapha and concha.

Use. To draw the ear back and stretch the concha.

Before describing the proper muscles of the ear, notice the following. The part usually called ear is for
the most part cartilaginous; appended to it below is a
pendulous portion, the LOBUS, composed of granulated
fat and dense cellular membrane; with the exception of
this, all above is the PINNA. The deep cavity in the
middle is the CONCHA. Dividing the upper part of the
concha into two unequal fossæ is a ridge which runs
along the circumference of the pinna and ends in the
lobus; this is the HELIX—within it another, the ANTI-
HELIX. On the anterior part of the concha is a projec-

tion, the TRAGUS, and opposite another, the ANTITRA-
GUS. Under the rim of the helix is the FOSSA INNOMI-
NATA, and in the bifurcation of the helix is the SCAPHA.
The external AUDITORY MEATUS is about one inch in
depth, its direction moderately forward and inward;
under the skin which lines the meatus are the CERU-
MINOUS GLANDS.

2. The proper muscles of the ear must be here described; but the student must not expect to meet with
them distinctly marked in every subject: in general,
they are very confused and indistinct.

(1) HELICIS MAJOR — *Arises* from the upper and
acute part of the helix, anteriorly.

Inserted into its cartilage, a little above the tragus.

Use. To depress the part from which it arises.

(2) HELICIS MINOR — *Arises* from the inferior and
anterior part of the helix.

Inserted into the crus of the helix, near the fissure on
the cartilage opposite to the concha.

Use. To contract the fissure.

(3) TRAGICUS — *Arises* from the middle and outer
part of the concha, at the root of the tragus, along
which it runs.

Inserted into the point of the tragus.

Use. To pull the point of the tragus a little forward.

(4) ANTITRAGICUS — *Arises* from the internal part of
the cartilage that supports the antitragus; and, running
upward, is

Inserted into the tip of the antitragus as far as the
inferior part of the antihelix.

Use. To turn the tip of the antitragus a little outward, and depress the extremity of the antihelix toward it.

(5) TRANSVERSUS AURIS — *Arises* from the prominent
part of the concha on the dorsum of the ear.

Inserted opposite to the outer side of the antihelix.

Use. It draws the parts to which it is connected toward each other, and stretches the scapha and concha.

CHAPTER III.

DISSECTION OF THE FACE.

Of the Muscles.

To expose these muscles an incision should be carried from the root of the nose to its tip, around the margin of the nostril to the centre of the upper lip, down to its margin, around its border to an opposite point on the lower lip, and then to the symphysis of the chin. From the termination of the last a second may be carried along the base of the jaw to the mastoid process of the temporal bone. Reflect the integuments outward.

Under the integuments of the face, there is always a considerable quantity of adipose membrane; many of the muscles are very slender, and lying imbedded in this fat, require careful dissection. The whole side of the face is also supplied with numerous ramifications of the facial nerve, or portio dura of the seventh pair. These nervous twigs are generally removed with the integuments. If it is desired to save them, the skin only should be removed first.

Twelve pairs of muscles, and one single muscle, are described in this dissection.

1. The ORBICULARIS PALPEBRARUM—*Arises* from the internal angular process of the frontal bone, and from a tendon at the inner angle of the eye, by a number of fleshy fibres which pass round the orbit, covering first the superior and then the inferior eyelid, and also the bony edges of the orbit.

Inserted, by a short round tendon, the TENDO OCULI, into the nasal process of the superior maxillary bone. The above tendon lies across the lachrymal sac.

Situation. This muscle is intermixed, at its upper part, with the occipito-frontalis.

Use. To shut the eye, by bringing down the upper lid, and pulling up the lower; the fibres contracting toward the inner angle, as to a fixed point, compress the eyeball and lachrymal gland, and convey the tears toward the puncta lachrymalia.

The ciliaris is only a part of the muscle covering the cartilages of the eyelids, which are called the Cilia or Tarsi; Maxillo-palpebral.

2. The CORRUGATOR SUPERCILII—*Arises*, fleshy, from the internal angular process of the os frontis; it runs outward and a little upward, to be

Inserted into the inferior fleshy part of the occipito-frontalis muscle, extending outward as far as the middle of the superciliary ridge.

Situation. This muscle is concealed by the occipito-frontalis. It lies close to the upper and inner part of the orbicularis palpebrarum, with which it is connected.

Use. To smooth the skin of the forehead, by pulling it down after the action of the occipito-frontalis. When it acts more forcibly, it pulls down the eyebrow and skin of the forehead, and produces vertical wrinkles.

3. The COMPRESSOR NARIS—*Arises*, narrow, from the outer part of the ala nasi, and neighboring part of the os maxillare superius. From this origin a number of thin separate fibres run up obliquely along the cartilage of the nose toward the dorsum nasi, where the muscle joins its fellow, and is

Inserted, slightly, into the lower part of the os nasi and nasal process of the superior maxillary bone.

Situation. It is superficial; its origin is connected with the levator labii superioris alæque nasi.

Use. To compress the ala toward the septum nasi; but, if the fibres of the occipito-frontalis, which adhere to it, act, the upper part of this muscle assists in pulling the ala outward. It also corrugates the skin of the nose.

4. LEVATOR LABII SUPERIORIS ALÆQUE NASI—*Arises* by two distinct origins: the first from the nasal process of the superior maxillary bone, where it joins the os frontis at the inner canthus of the eye; it descends along

the nasal process, and is *inserted* into the outer part of the ala nasi, and into the upper lip. The second *arises*, broad and fleshy, from the margin of the orbitar process of the superior maxillary bone, immediately above the foramen infra-orbitarium; it runs down, becoming narrower, and is *inserted* into the upper lip and orbicularis oris.

Situation. The first portion is sometimes called Levator Labii Superioris Alæque Nasi, and the second Levator Labii Superioris Proprius. Their origins are partly covered by the orbicularis palpebrarum. They descend more outwardly than the ala nasi.

Use. To raise the upper lip toward the orbit, and a little outward; the first portion will also draw the ala nasi upward and outward.

BLOODVESSELS AND NERVES.—Below the orbital origin of this muscle will be seen the INFRA-ORBITAL BLOODVESSELS AND NERVES coming out of the infraórbital foramen. The artery is from the internal maxillary, the nerve from the superior maxillary branch of the fifth pair; a nerve of common sensation.

5. ZYGOMATICUS MINOR — *Arises* from the upper prominent part of the os malæ, and, descending obliquely downward and forward, is

Inserted into the upper lip near the corner of the mouth.

Situation. Its origin is covered by the orbicularis palpebrarum.

Use. To draw the corner of the mouth and upper lip obliquely upward and outward.

6. ZYGOMATICUS MAJOR—*Arises*, fleshy, from the os malæ, near the zygomatic suture.

Inserted into the angle of the mouth, appearing to be lost in the depressor anguli oris, and orbicularis oris.

Situation. Its origin is partially covered by the orbicularis palpebrarum; it lies more outwardly than the zygomaticus minor.

Use. To draw the corner of the mouth and under lip upward and outward.

7. The LEVATOR ANGULI ORIS—*Arises*, thin and

fleshy, from a depression of the superior maxillary bone betwixt the root of the socket of the first dens molaris, and the foramen infra-orbitarium.

Inserted, narrow, into the angle of the mouth.

Situation. It lies more outwardly than the levator labii superioris alæque nasi; it is in part concealed by that muscle, by the zygomaticus minor, and part of the zygomaticus major. At its insertion it is connected with the depressor anguli oris.

Use. To draw the corner of the mouth upward.

8. The DEPRESSOR ANGULI ORIS—*Arises*, broad and fleshy, from the lower edge of the inferior maxillary bone, at the side of the chin, and gradually becoming narrower, is

Inserted into the angle of the mouth.

Situation. This muscle is firmly connected with the platysma myoides; at its insertion it is blended with the zygomaticus major and levator anguli oris.

Use. To pull down the corner of the mouth.

9. The DEPRESSOR LABII INFERIORIS—*Arises*, fleshy and broad, from the side of the lower jaw, a little above its lower edge; it runs obliquely upward and inward, and is

Inserted into the edge of the under lip.

Situation. This muscle, at its insertion, decussates with its fellow. It is in part covered by the depressor anguli oris. It forms the thick part of the chin, and has its fibres interwoven with fat.

Use. To pull the under lip downward.

10. The BUCCINATOR—*Arises*, tendinous and fleshy, from the lower jaw as far back as the root of the coronoid process; from the upper jaw, as far back as the pterygoid process of the sphenoid bone; it then continues to arise from the alveolar processes of both jaws, as far forward as the dentes cuspidati. The fibres run forward, and are

Inserted into the angle of the mouth.

Situation. This muscle lies deep, adheres to the membrane that lines the mouth; and a quantity of fat is always found between its fibres and the other muscles

Fig. 1.

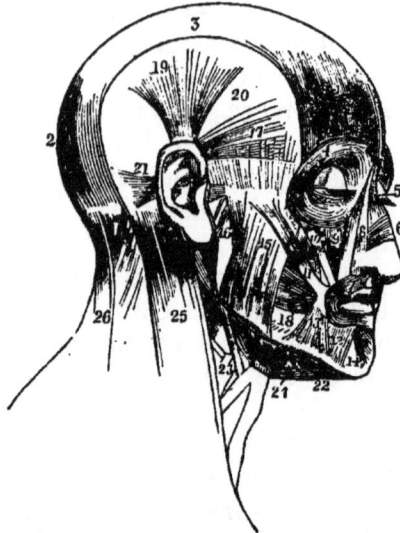

MUSCLES OF THE FACE.

1. Frontal portion of the Occipito-Frontalis.
2. Its Posterior or Occipital portion.
3. Its Aponeurosis.
4. Orbicularis Palpebræ, which
conceals the Corrugator
Supercilli, and Tensor
Tarsi of Horner.
5. Pyramidalis Nasi.
6. Compressor Naris.
7. Orbicularis Oris.
8. Levator Labii Superioris
Alæque Nasi.
9. Levator Labii Superioris.
10. Zygomaticus Minor.
11. Zygomaticus Major.
12. Depressor Labii Inferioris.
13. Depressor Anguli Oris.
14. Levator Labii Inferioris.

15. Superficial Portion of the
Masseter.
16. Its Deep Portion.
17. Attrahens.
18. Buccinator.
19. Attollens Auriculæ Muscle.
20. Temporal Aponeurosis which
conceals the Temporal
Muscle.
21. Retrahens Auriculæ Muscle.
22. Anterior Belly of the Digastric—the Tendon is seen
passing through the Loop
formed by the Cervical
Fascia.
23. Stylo-hyoid Muscle.
24. Mylo-hyoid Muscle.
25. Sterno-mastoid Muscle.
26. Upper part of the Trapezius
—the Muscle between 25
and 26 is the Splenius.

and integuments. It is partly concealed by the masseter,
and by the muscles which pass to the angle of the mouth,
as the levator and depressor anguli oris, and zygomaticus
major. It is inserted behind these muscles. In the
cheek it is connected with the platysma myoides.

Use. To draw the angle of the mouth backward and outward, and to contract its cavity, by pressing the cheek inward.

The single muscle is the

ORBICULARIS ORIS.—It consists of two planes of semicircular fibres, which decussate at the angles of the mouth. These fibres are formed chiefly by the muscles which are inserted into the lips; they surround the mouth. The superior portion runs along the upper lip, the inferior portion along the under lip.

Situation. It is connected and intermixed with the insertions of all the preceding muscles of the face. Some of the fibres are connected to the septum nasi, and are by Albinus termed Nasalis Labii Superioris.

Use. To shut the mouth by contracting and drawing both lips together.

At this stage of the examination the FACIAL ARTERY will be brought into view, a branch of the external carotid mounting over the lower jaw in front of the masseter muscle, passing under the depressor anguli oris and ascending as high as the angle of the eye, where it is called ANGULAR ARTERY. Its branches, enumerated from below upward, are as follows: MASSETER branches to the masseter and buccinator muscles. INFERIOR LABIAL to muscles of the lower lip. INFERIOR CORONARY to edge of lower lip. SUPERIOR CORONARY along the edge of the upper lip. LATERALIS NASI to the ala and septum of the nose. NERVES.—ANTERIOR DENTAL, some distance back from the symphysis of the lower jaw, comes out of the anterior dental foramen, and is the termination of the inferior maxillary or 3d branch of the 5th pair. Any filaments coming forward from the outer portion of the face are from the portio dura of the 7th pair.

11. DEPRESSOR LABII SUPERIORIS ALÆQUE NASI— *Arises,* thin and fleshy, from the os maxillare superius, where it forms the alveoli of the dentes incisiva and dens caninus; thence it runs up under part of the levator labii superioris alæque nasi.

Inserted into the upper lip and root of the ala nasi.

Situation. It is concealed by the orbicularis oris and levator labii superioris alæque nasi. It may be discovered by inverting the upper lip, and dissecting on the side of the frænum, which connects the lip to the gums.

Use. To draw the upper lip and ala nasi downward and backward.

12. The LEVATOR LABII INFERIORIS—*Arises* from the lower jaw at the root of the alveolus of the lateral incisor.

Inserted into the under lip and the skin of the chin.

Situation. Those two small muscles are found by the side of the frænum of the lower lip. They lie under the depressor labii inferioris.

On the side of the face we observe two strong muscles, and two other muscles are concealed by the angle of the inferior maxilla.

Behind the masseter and overlaying a portion of it is the PAROTID GLAND (a salivary gland). It is wedged in between the angle and ramus of the jaw and the cartilage of the ear, extending up as high as the zygoma, and down into the neck. Its deep surface rests upon the styloid process of the temporal bone and is insinuated into all the irregularities about this point: its color is pinkish, and consists of lobules held together by cellular tissue. It has no proper capsule, but is covered by the extension of the fascia of the neck. The external carotid artery runs through its substance, giving it many branches, called PAROTIDEAN. The PORTIO DURA, or facial nerve, after it escapes from the stylomastoid foramen, passes through it also on its way to the face, and its several branches form a looped connection over the masseter called the PES ANSERINUS. From the anterior and upper part, a white canal (sometimes two for a little way) passes forward over the masseter muscle to its anterior edge, beneath which it dips, and, perforating the buccinator, empties into the mouth opposite the 2d molar of the upper jaw. This is the PAROTID DUCT (duct of Steno). It is accompanied by branches of the portio dura nerve, and the TRANSVERSE FACIAL ARTERY from the external carotid.

1. The MASSETER is divided into two portions, which decussate one another.

The Anterior Portion *arises*, tendinous and fleshy, from the superior maxillary bone, where it joins the os malæ; from the lower edge of the os malæ, and from its zygomatic process. The strong fibres run obliquely downward and backward, and are *inserted* into the outer surface of the side of the lower jaw, extending as far back as its angle.

The Posterior Portion *arises*, principally fleshy, from the inferior surface of the os malæ, and of the whole of the zygomatic process, as far back as the tubercle before the socket for the condyle of the lower jaw. The fibres slant forward, and are *inserted*, tendinous, into the outer surface of the coronoid process of the lower jaw.

Situation. The anterior portion conceals almost the whole of the posterior portion. The greater part of this muscle is superficial. Below it is covered by the platysma myoides; and above, a small portion of it is concealed by the origin of the zygomaticus major.

Use. To pull the lower to the upper jaw, and to move it forward and backward.

2. TEMPORALIS—*Arises*, fleshy, from a semicircular ridge in the lower and lateral part of the parietal bone, from all the squamous portion of the temporal bone, from the external angular process of the os frontis, from the temporal process of the sphenoid bone, and from an aponeurosis which covers the muscle. From these different origins the fibres converge, descend under the bony jugum formed by the zygomatic processes of the temporal and cheek bones.

Inserted, by a strong tendon, into the upper part of the coronoid process of the lower jaw, to which it adheres on every side, but more particularly its forepart, where the insertion is continued down to near the last dens molaris. If the zygoma is removed, a better view will be had.

Situation. This muscle is of a semicircular shape. It is covered by a fascia or aponeurosis. This fascia adheres to the bones which give origin to the upper part of the muscle, and descending over it, is inserted into the jugum and adjoining part of the os malæ and os frontis. This insertion is by two leaves, leaving a space between containing some fat. The temporalis, at its origin, lies under the expanded tendon of the occipito-frontalis, and under the small tendons which move the external ear. Its insertion is concealed by the jugum and by the masseter; so that, to expose it, the masseter must be cut away.

Use. To pull the lower jaw upward and press it against the upper.

Fig. 2.

THE TWO PTERYGOID MUSCLES. THE ZYGOMATIC ARCH AND THE GREATER PART OF THE RAMUS OF THE LOWER JAW HAVE BEEN REMOVED, IN ORDER TO BRING THESE MUSCLES INTO VIEW.

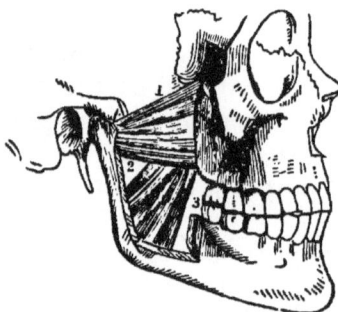

1. The Sphenoid Origin of the External Pterygoid Muscle.
2. Its Pterygoid Origin.
3. The Internal Pterygoid Muscle.

In order to expose the following muscles, we must re-move the muscles of the cheek and jaw, the masseter and insertion of the temporalis must be taken away, and the coronoid process of the inferior maxilla removed by a saw.

3. The PTERYGOIDEUS EXTERNUS—*Arises* from the outer side of the external plate of the pterygoid process of the sphenoid bone, from part of the tuberosity of the os maxillare adjoining to it, and from the root of the temporal process of the sphenoid bone. It passes back-ward and outward, to be

Inserted into a depression in the neck of the condy-loid process of the lower jaw, and into the anterior and inner part of the ligament of the articulation of that bone.

Situation. This muscle passes almost transversely from the skull to its insertion. It is concealed by the muscles of the face and neck, and by the ascending processes of the lower jaw.

Use. When this pair of muscles act together, they bring the jaw horizontally forward. When they act singly, the jaw is moved forward, and to the opposite side.

4. The PTERYGOIDEUS INTERNUS—*Arises*, tendinous and fleshy, from the inner and upper part of the internal plate of the pterygoid process of the sphenoid bone, filling all the space between the two plates; and from the pterygoid process of the os palati between these plates.

Inserted, by tendinous and fleshy fibres, into the inside of the angle of the lower jaw.

Situation. To expose this muscle, the jaw must be removed from its articulating cavity, and then pulled forward, and toward the opposite side; or it may be sawn across at its symphysis, and the other half removed. It is larger than the pterygoideus externus; and between the two muscles there is a considerable quantity of cellular membrane, and the trunk of the INFERIOR MAXILLARY and GUSTATORY NERVES. Like that muscle, it is concealed by the lower jaw and facial muscles. Along its posterior edge, we observe the Ligamentum Laterale Maxillæ Inferioris, a ligamentous band, which extends from the back part of the styloid process to the angle of the lower jaw.

Use. To draw the jaw upward, and obliquely toward the opposite side.

Arteries.

The External Carotid is found ascending behind the parotid gland. It perforates the gland at its upper part, ascends over the zygomatic process immediately before the ear, and divides into the anterior, middle, and posterior temporal arteries, which ramify over the side of the head, giving also branches to the forehead and occiput.

The INTERNAL MAXILLARY passes behind the condyloid process of the lower jaw; it directs its course toward the bottom of the orbit of the eye; and it is at this point that it sends off its numerous branches. (1) ARTERIA MEDIA DURÆ MATRIS passes through the spinal hole of the sphenoid bone into the cranium, and is distributed to the dura mater. (2) A. MAXILLARIS INFERIOR, or Inferior Dental, runs downward, enters the foramen at the root of the ascending processes of the lower jaw, then passes through the canal of the lower jaw; supplying the teeth and sockets, and emerges by the foramen mentale, to be distributed to the chin. It is accompanied by a nerve and one or two veins. (3) A. PTERYGOIDEÆ and A. TEMPORALES PROFUNDÆ are small branches of the internal maxillary which pass to the pterygoid muscles, and to the inner part of the temporal muscle. (4) A. PHARYNGEÆ, branches to the pharynx, palate, and base of the skull. (5) A. ALVEOLARIS, which gives branches to the teeth of the upper jaw, and to the jaw bone itself. (6) A Branch (spheno-palatine) through the foramen spheno-palatinum to the nose; and (7) An Artery through the palato-maxillary canal to the palate.

Fig. 3.

THE INTERNAL MAXILLARY ARTERY AND ITS BRANCHES.

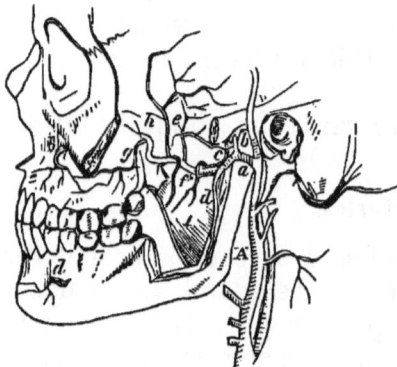

A. External Carotid Artery.
a. Internal Maxillary Artery.
b. Arteria Tympanica.
c. Arteria Pterygoidea.
d. Dentalis Inferior.
e. Arteria Meningea Parva.
f. Arteria Buccalis.
g. Arteria Alveolaris, or Maxillaris Superior.
h. Arteria Meningea Magna.
o. Infra-Orbitalis.
d. Anterior Mental Artery.

The continued trunk of the internal maxillary enters the orbit by the spheno-maxillary slit. It sends off a branch which runs along the inner side of the orbit, and passes out at the inner canthus of the eye on the forehead. The artery itself runs along the bottom of the orbit in a canal on the upper part of the great tuberosity of the os maxillare superius, and emerges by the foramen infra-orbitarium on the face; hence it is termed A. Infra-Orbitaria, and is distributed to the cheek and side of the nose.

The FRONTAL ARTERY is also seen in the dissection of the face, passing from the orbit through the foramen supra-orbitarium to be distributed to the forehead. This artery is sent off from the ophthalmic artery, which is a branch of the internal carotid.

If the face be injected, a remarkable anastomosis of arteries will be observed at the inner angle of the eye.

Veins.

The veins of the face are numerous, and pass into the external and internal jugular veins.

Nerves.

1. The PORTIO DURA of the seventh pair, Nervus Communicans Faciei, or Facial nerve, after its course through the temporal bone in the aqueduct of Fallopius, comes out by the foramen stylo-mastoideum. It immediately gives off branches to the neighboring parts, as behind the ear. It then passes through the substance of the parotid gland, and emerges on the face in three great branches, which have frequent mutual communications: this division of the nerve is called PES ANSERINUS.

(1) The ascending branch ramifies on the temple and forehead.

(2) The middle branch sends its ramifications over the side of the face, the proper Facial Nerves.

(3) The descending branch sends its twigs along the chin, down upon the neck, and backward upon the occiput.

2. The SUPERIOR CERVICAL NERVES send off several

Fig. 4.

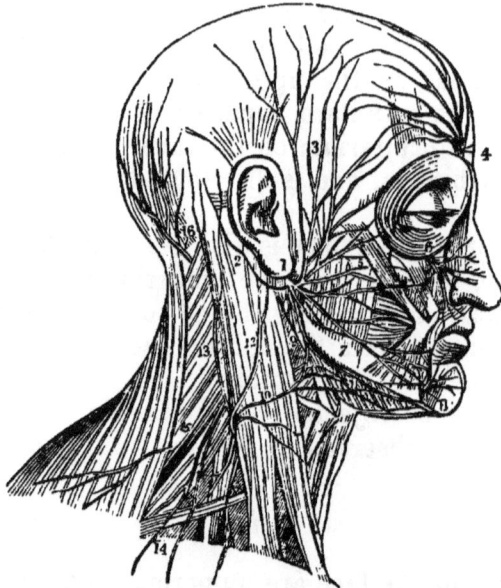

THE DISTRIBUTION OF THE FACIAL NERVE AND THE BRANCHES OF
THE CERVICAL PLEXUS.

1. The Facial Nerve (Portio Dura) escaping from the Stylo-mastoid
   Foramen, and crossing the Ramus of the Lower Jaw; the
   Parotid Gland has been removed, in order to show the Nerve
   more distinctly.
2. The posterior Auricular Branch; the Digastric and Stylo-mastoid
   Filaments are seen near the origin of this branch.
3. Temporal Branches communicating with (4) the Branches of the
   Frontal Nerve.
5. Facial Branches communicating with (6) the Infra-orbital Nerve.
7. Facial Branches communicating with (8) the Mental Nerve.
9. Cervico-facial Branches communicating with (10) the Superficial
   Cervical Nerve, and forming a Plexus (11) over the Submaxillary Gland. The distribution of the branches of the facial
   in a radiated direction over the side of the face, and their
   looped communications, constitute the Pes Anserinus.
12. The large Auricular Nerve, one of the Ascending Branches of
    the Cervical Plexus.
13. The Small Occipital ascending along the Posterior Border of the
    Sterno-mastoid Muscle.
14. The Superficial and Deep-descending Branches of the Cervical
    Plexus.
15. The Spinal Accessory Nerve, giving off a Branch to the External
    Surface of the Trapezius Muscle.
16. The Large Occipital Nerve, the Posterior Branch of the second
    Cervical Nerve.

branches, which ramify on the side of the face and head, and communicate freely with the branches of the portio dura.

As, in the course of this dissection, we meet with many twigs of the second and third branches of the fifth pair of nerves, it will be advisable here to describe this pair of nerves.

1st. The OPHTHALMIC NERVE. The first branch of the fifth pair enters the orbit through the sphenoidal fissure. Its branches are the FRONTAL, which, passing above the levator palpebral muscles, escapes upon the forehead through the supra-orbital foramen. The LACHRYMAL passes along the upper edge of the external rectus muscle with the LACHRYMAL ARTERY to the LACHRYMAL GLAND. Here it divides into two branches, one escaping through the malar bone upon the cheek and temple, the other to the under part of the gland, and conjunctiva of the upper lid.

The NASAL passes between the origins of the external rectus muscle, accompanies the ophthalmic artery and enters the anterior ethmoidal foramen, then up through the ethmoid bone to the cribriform plate, and then through the nasal slit in that plate into the nose, which supplies the mucous membrane and integuments of the nose. Its principal branches while in the orbit are the GANGLIONIC, CILIARY, and infra-trochlear.

3. NERVUS MAXILLARIS SUPERIOR, or second branch of the fifth pair. The superior maxillary nerve, having left the cranium by the foramen rotundum of the sphenoid bone, emerges behind the antrum maxillare, at the lower back part of the orbit, and at the root of the ptery-goid process of the sphenoid bone. It immediately sends out branches: 1. A small branch which passes through the spheno-maxillary slit to the periosteum and fat of the orbit. 2. The largest branch is the INFRA-ORBITARY NERVE. It enters the channel in the top of the antrum maxillare, accompanying the infra-orbitary artery, comes out at the foramen infra-orbitarium, and is widely distributed to the cheek, under lip, and outside of the nose, communicating with ramifications of the portio dura.

3. Branches to the temporal muscle, os malæ, etc. 4. Communicating branches with Meckel's ganglion. 5. Posterior dental branches through the foramina in the superior maxillary bone. 6. Twigs which supply the gums and alveoli of the upper jaw.

4. NERVUS MAXILLARIS INFERIOR, or third branch of the fifth pair. The inferior maxillary nerve leaves the cranium by the foramen ovale of the sphenoid bone. It has its course downward and outward; and, having given twigs to the parts near which it passes, as the masseter, pterygoid, and temporal muscles, it divides at the angle of the jaw into two branches.

1. INFERIOR DENTAL NERVE—Enters the foramen at the angle of the lower jaw, accompanies the INFERIOR DENTAL ARTERY along the canal in that bone, giving nerves to the teeth, emerges at the anterior mental foramen, and is distributed to the chin.

2. The GUSTATORY or LINGUAL NERVE passes to the tongue. It lies close along the inner surface of the lower jaw, below the inferior edge of the pterygoideus-internus, and above the mylo-hyoid will be seen when the neck is dissected. This nerve receives the chorda tympani.

3. The FRONTAL NERVE, which comes from the first branch of the fifth pair through the supra-orbital foramen: it is distributed to the forehead.

A general view of the distribution of the spinal nerves may be usefully subjoined in this place, to complete the description of the nerves.

1. The CERVICAL NERVES consist of eight pairs; they spread their branches over the side and back of the neck and head, and to the muscles moving the head and shoulders. The superior nerves send branches to the sides of the head, and the inferior to the upper part of the chest and back. They also communicate freely with each other, and with all the neighboring nerves;—high in the neck, and under the jaw, with the portio dura of the seventh pair, with the fifth, eighth, and ninth pairs, and with the great sympathetic;—toward the middle of the neck, with the descendens noni, the sympathetic and eighth pair, and in the lower part of the neck with the

The ˈPHRENIC NERVE is formed by branches of the third, fourth, and fifth cervical nerves, passes obliquely down the neck through the thorax, then on each side of the pericardium, and is distributed to the diaphragm.

The AXILLARY PLEXUS is formed by the principal parts of the trunks of the fourth, fifth, sixth, and seventh cervical, and first dorsal nerves.

2. The DORSAL NERVES are twelve pairs. They arise from the spinal marrow in the same manner as the cervical. Each nerve emerges betwixt the heads of the ribs, gives twigs to the great sympathetic nerve, and twigs which pierce backward to the muscles of the back; then entering the groove in the lower edge of each rib, it accompanies the intercostal artery, and runs toward the anterior part of the thorax, supplying the great muscles of the chest, giving twigs to the diaphragm, and muscles of the abdomen.

3. The LUMBAR NERVES are five pairs. They arise in the same manner; their trunks are covered by the psoas magnus muscle. Each nerve gives twigs to the muscles of the loins and back, and to the sympathetic nerves, and runs obliquely downward to supply the abdominal muscles and integuments of the groin and scrotum; but the trunks of these nerves assist in forming the nerves of the thigh.

4. The SACRAL NERVES are five on each side, arising from the cauda equina. They come out through the anterior foramina, and send small branches to the neighboring parts; but the great trunks of these nerves are united with the lumbar nerves, to form the nerves of the lower extremity, viz.:

(1) The ANTERIOR CRURAL NERVE, passing out under Poupart's ligament to the extensor muscles of the leg, is formed by branches of the first, second, third, and fourth lumbar nerves.

(2) The OBTURATOR NERVE, leaving the pelvis by the thyroid holes, and being distributed to the deep-seated muscles on the inside of the thigh, arises from branches of the second, third, and fourth lumbar nerves.

(3) The ISCHIATIC or SCIATIC NERVE, the greatest

nerve of the body, passes out from the back part of the
pelvis, through the sacro-sciatic notch, and takes its
course along the back of the thigh, to supply the thigh,
leg, and foot; it is formed from the two last nerves of
the loins and three first of the sacrum.

All these nerves of the spine communicate freely by
numerous twigs, and by the intervention of

The GREAT SYMPATHETIC NERVE, or INTERCOSTAL.—
This nerve, consisting of ganglia connected by cords,
passes out of the cranium with the carotid artery. It
then descends through the neck, and forms three ganglions in its course, which give twigs to the neighboring
parts, and are joined by filaments from the cervical
nerves, and the eighth and ninth pairs. The intercostal
then enters the thorax, and descends by the side of the
vertebra, behind the pleura, giving filaments, which,
joining with twigs of the eighth pair, form several plexuses to supply the heart, lungs, etc. In the abdomen it
descends on the lumbar vertebræ, and at last terminates
in the pelvis on the extremity of the coccyx.

While in the thorax, it gives off a branch, which,
uniting with branches of the dorsal nerves, forms

The ANTERIOR INTERCOSTAL, or SPLANCHNIC NERVE.
—This nerve, passing betwixt the crura of the diaphragm, enters the abdomen, forms the semilunar ganglion, and is distributed by numerous plexuses to all the
abdominal viscera.

The eighth pair, or par vagum, has also a very long
course; it arises in the head, passes through the neck, to
which it gives several branches. It enters the thorax
anterior to the subclavian artery; here it gives off a
remarkable branch, called the Recurrent, because it is
reflected round the arch of the aorta on the left side,
and round the subclavian artery on the right, and ascends to be distributed on the trachea, œsophagus, and
larynx. The nerve then passes through the thorax, and,
entering the abdomen, terminates in the stomach; in
this course it has frequent communications with the
great sympathetic, which it assists in forming the different plexuses that supply the thoracic and abdominal

viscera. It is distributed to the heart, lungs, liver, spleen, stomach, and duodenum.

The Nose.

The two openings in front are the NOSTRILS guarded by stiff hairs (VIBRISSÆ). The septum between the nostrils, the COLUMNA; the tip of the nose, the LOBULUS; sides (ALÆ). It is covered with great numbers of sebaceous follicles: its form secured by five fibro-cartilages.

1. FIBRO-CARTILAGE OF THE SEPTUM divides partly the nasal fossæ into two.

2. LATERAL FIBRO-CARTILAGE (2) connected above to the nasal bones and nasal process of superior maxillary, and below to the alar fibro-cartilage.

3. ALAR FIBRO-CARTILAGES (two); each is curved to correspond to the nostril. The inner portion of each

Fig. 5.

THE NASAL CARTILAGES, SHOWING THEIR CONNECTION WITH EACH OTHER AND WITH THE OSSI NASI.

1. Cartilage of the Septum.
2, 2. Lateral Cartilages.
3. Ala Cartilage.
4. Cornua, or Appendices of the Ala Cartilage.
5. Nostril.

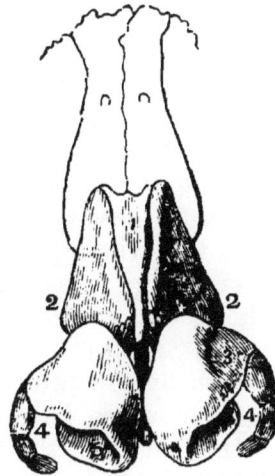

coming together form the columna. The alæ are extended by three or four fibro-cartilaginous appendages.

CHAPTER IV.

OF THE CONTENTS OF THE CRANIUM, OR THE BRAIN AND ITS MEMBRANES.

A TRANSVERSE incision, extending from ear to ear over the crown of the head, being made through the tendon of the occipito-frontalis, the two flaps may, with facility, be inverted on the face and neck. Remove the superior part of the cranium by a saw directed anteriorly through the frontal bone above the orbitar process, and posteriorly as low as the transverse ridge of the occipital bone. Thus the subsequent demonstration of the brain will be conducted with greater facility.

When the superior part of the cranium, commonly called the Calvarium, or skull cap, is torn off, which requires considerable force, you expose the DURA MATER, a firm, compact, and whitish membrane, somewhat shining, rough on its outer surface, from the rupture of vessels which connected it to the cranium, and covered with bloody spots in consequence of the blood effused from these ruptured orifices. It is described as being separable into many laminæ (into two with facility); and it is said that these two laminæ, by separating and reuniting, form the triangular cavities named Sinuses, which are in fact large veins. This division of layers can hardly be admitted as correct in the recent state of the membrane.

The SUPERIOR LONGITUDINAL SINUS lies in a groove formed by the two parietal bones; it extends along the sagittal suture from the crista galli of the ethmoid bone to the middle of the os occipitis, where it bifurcates into the two lateral sinuses; in its passage backward, its size is increased. When slit open (which it should be), its triangular form is evident; it is lined by a smooth membrane, and in it may be remarked the numerous openings of the veins of the pia mater opening forward, the

frena, or slips of fibres crossing from side to side, called
Chordæ Willisii, glandulæ Pacchioni internæ, et externæ, little bodies like millet-seed seen on the outer
and inner surface of the sinus.

Fig. 6.

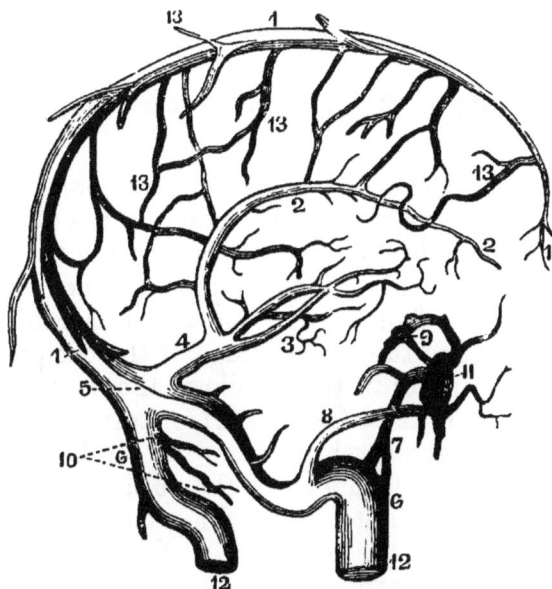

SINUSES OF THE DURA MATER.

1. Superior Longitudinal Sinus.
2. Inferior Longitudinal Sinus.
3. The two Venæ Galeni.
4. Sinus Quartus.
5. Torcular Herophili.
6, 6. The Lateral Sinuses.
7. Inferior Petrous Sinus.
8. Superior Petrous Sinus.
9. Circular Sinus of Ridley.
10. The two occipital Sinuses.
11. Cavernous Sinus.
12. Internal Jugular Veins.
13. Veins of the Pia Mater.

The arteries of the dura mater are divided into the
anterior, middle, and posterior.

1. Arteria Meningea Media (called also the Spinalis or Spheno-spinalis), the great middle artery, is a
branch of the internal maxillary; it passes through the
spinous hole of the sphenoid bone, and is seen arising
from the anterior inferior angle of the parietal bone (in
a groove on which it lies), and spreading its numerous
branches over the dura mater.

3

The anterior and posterior arteries are small.

2. A. MENINGEA ANTERIOR is sent off from the external carotid, and enters the cranium by the foramen lacerum orbitale superius.

3. A. MENINGEA POSTERIOR is given off by the vertebral artery; the dura mater also receives small twigs from the occipital, pharyngeal arteries, etc.

Of the Septa of the Brain, or Processes of the Dura Mater.

1. The FALX (cerebri or major) is a long and broad fold, or duplicature of the inner lamina of the dura mater, dividing the cerebrum into two hemispheres, extending from the crista galli of the ethmoid bone, along the middle of the os frontis and point of junction of the two parietal bones, to the crucial ridge of the occipital bone, where it terminates in the middle of the next septum.

2. The TENTORIUM CEREBELLI, or transverse septum. This separates the cerebrum from the cerebellum, and is formed by the inner lamina of the dura mater, reflected off from the os occipitis along the groove of the lateral sinuses, and the edge or angle of the temporal bones. Its position is horizontal.

There are some other folds of the dura mater not visible in this stage of the dissection.

3. The falx of the cerebellum, or small occipital septum, will be seen when the cerebrum is removed. It extends from the middle of the tentorium along the middle spine of the os occipitis to the foramen magnum, dividing the cerebellum into two parts.

4. The sphenoidal folds, two small folds of the dura mater, one on each side of the sella turcica, stretching from the posterior to the anterior clinoid processes.

The dura mater, also, in many parts of the brain, separates its laminæ to form sinuses; the principal of these will be noticed in the course of the dissection. This membrane should now be divided in the line of the division of the cranium; its internal surface is smooth, glistening, and free from adhesion, except in the course of the longitudinal sinus; into which veins pass from the pia mater.

Fig. 7.

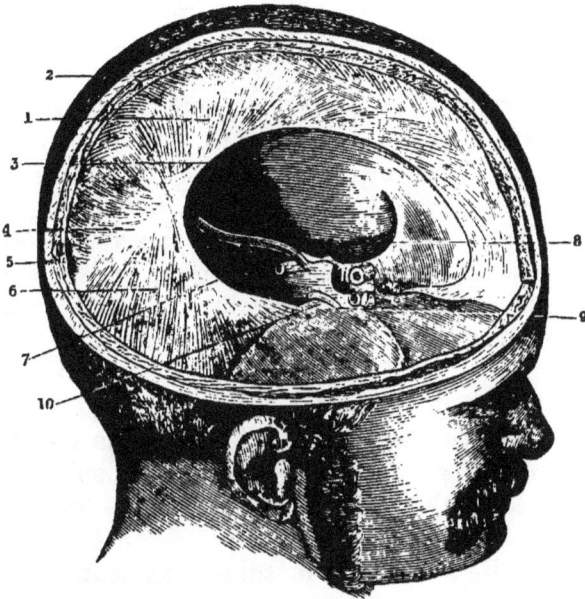

OBLIQUE VIEW OF THE INTERIOR OF THE CRANIUM AS LINED BY THE DURA MATER.

1. Falciform Process.
2. Its Superior or Attached Border containing the Longitudinal Sinus.
3. Its Free Border.
4. Continuation of the Falciform Process with (6) the Tentorium.
7, 8. Free Concave Edge of the Tentorium.
9. Termination of this edge at the Anterior Clinoid Process.
10. Attached Border of the Tentorium continued along the Upper Angle of the Petrous Bone to the Posterior Clinoid Process.

Detach the falx from the crista galli, and turn it backward, observe in its lower edge the INFERIOR LONGITUDINAL SINUS, which enters a sinus in the tentorium, termed the STRAIGHT SINUS. This will fully expose the convolutions of the brain, which are closely invested by the pia mater.

1. The TUNICA ARACHNOIDES is a fine membrane, covering uniformly the surface of the pia mater, with-

out passing into the interstices of its duplicatures.   It
is attached to it, is extremely thin, transparent, without
vessels, demonstrated with difficulty on the upper surface of the brain by the blowpipe (which raises it into
cells), but on the base of the brain it can be distinctly
seen.

2. The proper Pia Mater, or tunica vasculosa, is a
very vascular membrane, transparent in the interstices
of its vessels, investing the substance of the brain, descending betwixt all its convolutions, and lining its different cavities; but, where it lines the ventricles, it is
fine, delicate, and less vascular than on the surface and
betwixt the convolutions of the brain.   It is connected
to the dura mater by its veins passing into the longitudinal sinus.

The brain is divided into three parts: 1. The cerebrum; 2. The cerebellum; 3. The medulla oblongata.

The CEREBRUM consists of two distinct substances:

1. The cortical or vesicular substance forming the
outer part.

2. The white medullary or tubular substance forming
the inner part.

The brain is divided by the falx into two hemispheres,
and by the pia mater into numerous convolutions.

Each hemisphere is divided into three lobes.

1. The ANTERIOR LOBES rest on that part of the cranium which forms the two orbits, and is called the anterior fossæ of the basis of the cranium.

2. The MIDDLE LOBES are situated before and above
the medulla oblongata, and rest on the middle fossæ of
the basis cranii, which are formed by the sphenoid and
temporal bones.

3. The POSTERIOR LOBES are supported by the tentorium.

The anterior and middle lobes are parted by a deep
narrow sulcus, which ascends obliquely backward from
the temporal ala of the os sphenoides to near the middle
of the os parietale; it is termed FISSURA CEREBRI, or
Fissura Magna Sylvii.

By gently separating with the fingers the two hemispheres of the brain,[1] we see passing betwixt them a longitudinal white convex body, the CORPUS CALLOSUM: it lies under the falx.   On the surface of the corpus cal-

Fig. 8.

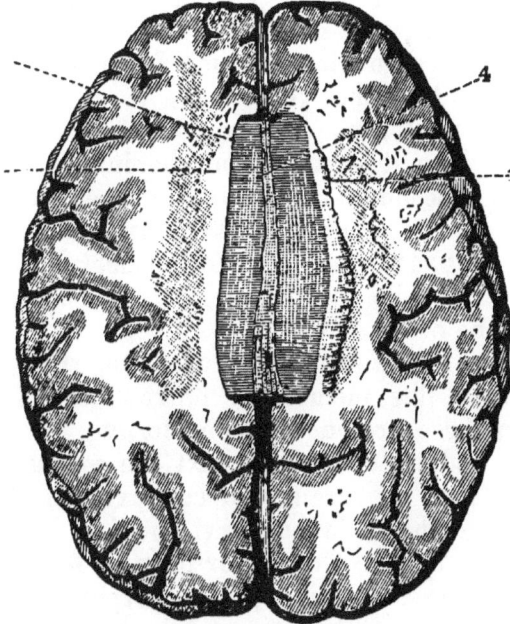

HORIZONTAL SECTION OF THE CEREBRUM UPON A LEVEL WITH THE
CORPUS CALLOSUM.

1. Outer edge of the Corpus Callosum, formed by pressing aside the
    Medullary Substance of the Hemisphere.
2. Medullary or Fibrous Substance.
3. Upper Surface of the Commissure.
4. Raphé.

losum is seen the RAPHÉ, between two longitudinal med-ullary lines, LINEA LONGITUDINALIS LANCISII, united by transverse fibres, the LINEA TRANSVERSA.

   When one hemisphere of the brain is cut horizontally

---

[1] Between the hemispheres and on the surface of the corpus callo-sum, we observe the arteriæ callosæ, which are the continuation of the trunks of the anterior cerebri.

on the level of the corpus callosum, an appearance is produced, termed the CENTRUM OVALE MINUS. When both are sliced off to the same level, the CENTRUM OVALE MAJUS.

Under this surface are the two lateral ventricles.[1] If one of these be cautiously perforated on the side of the corpus callosum, and gently inflated by a blowpipe, its extent may be seen; but if much force be used, the air will pass into the other ventricle.

The two ventricles are separated by a medullary partition, which descends from the inferior surface of the corpus callosum to the fornix, the SEPTUM LUCIDUM; it consists of two laminæ, with a narrow cavity between the FIFTH VENTRICLE. To see this septum, one of the ventricles must be laid open, and the septum pulled gently to the other side.

The LATERAL VENTRICLES are two, right and left, lined with a fine membrane, narrow, consisting of a body, and three prolongations or cornua.

1. The body is formed betwixt the corpus callosum, the medulla of the brain, the convexity of the corpus striatum, and the thalamus nervi optici.

2. The ANTERIOR CORNU, or horn, is formed betwixt the more acute convexity of the corpus striatum and the anterior part of the corpus callosum.

3. The POSTERIOR CORNU (called also the digital cavity) may be traced stretching backward and downward into the posterior lobe of the brain.

4. The inferior or descending cornu cannot be traced in this stage of the dissection; it seems like the continued cavity of the ventricle, takes a curve backward and outward, and then, turning forward, descends into the middle lobe of the brain.

The lateral ventricles communicate with each other,

[1] To show the lateral ventricles, the corpus callosum should be cut away close to the septum lucidum, and then the ventricle of that body, and the thickness and breadth of the septum itself, will be more clearly seen.

and with the third ventricle, by an opening under the forepart of the arch of the fornix.[1]

In the lateral ventricles we meet with

The FORNIX, a medullary body, flat and of a triangular shape, broadest behind, which divides the two lateral and the third ventricles. It is exposed on tearing away the septum lucidum; its lower surface is toward the third ventricle; its lateral margins are in the lateral ventricles; on its upper surface it supports the septum lucidum, and under its most anterior part is the foramen Monroianum. One of the angles of this body is forward, and the other two toward the back part: it rests chiefly on the thalami nervorum opticorum, but is separated from them by a vascular membrane called the velum interpositum.

The extremities of the body of the fornix are named its Crura.

1. The crus anterius is double, bends downward before the anterior commissure of the brain, with which it is connected, and may be traced into the corpora albicantia, and tuber cinereum on the base of brain.

2. The two crura posteriora, coalescing with the back part of the corpus callosum, pass, on each side, into the inferior cornu of the lateral ventricle, and terminate on the hippocampus major.

Divide the body of the fornix, invert it, by turning the anterior crus forward, and the posterior crura backward; on the under surface of the latter is an appearance of transverse lines, named Corpus Psalloides, psalterium, or lyra.

The inversion of the fornix exposes

---

[1] It has been doubted whether or not this be an opening: the choroid plexus passes through it, and seems to unite the surfaces; it is absurdly named the Foramen Monroianum, from a mistaken notion that Dr. Munro discovered it, and may be seen by gently turning the anterior crus of the fornix to one side; it is a space betwixt the most anterior part of the convexity of the thalami nervorum opticorum, and the anterior crus of the fornix.

This foramen may always be easily found by following the course of the plexus choroides, as it passes forward in the ventricle. It is a slit, rather than a round hole, in the natural state.

The Plexus Choroides.—This is a continuation of the pia mater, a spongy mass, consisting of folds of tor-

Fig. 9.

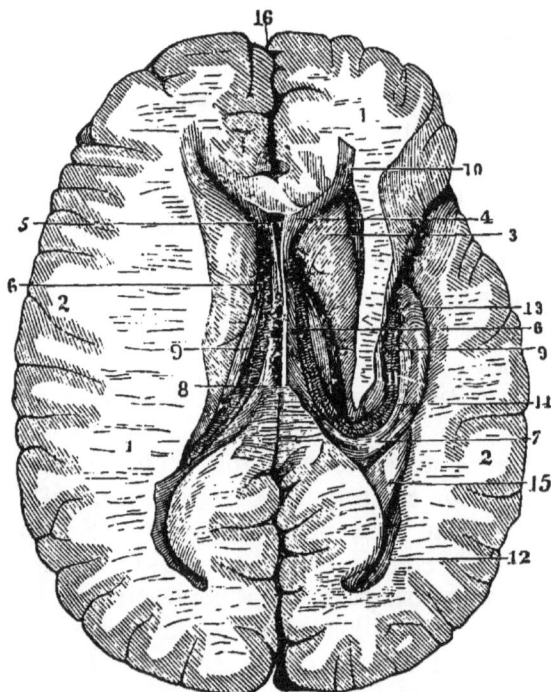

A Transverse Section of the Brain, showing the Corpora Striata, Lateral Ventricles, and the Associated Parts.

1, 1. Medullary portion of the Hemispheres.
2. Vesicular Neurine or Cortical Portion.
3. Corpus Striatum.
4. Septum Lucidum.
5. Ventriculus Septi or Fifth Ventricle.
6, 6. The Fornix.
7. Posterior Crura of the Fornix.
8. Base of the Fornix.
9, 9. Plexus Choroides, at the Margin of the Velum Interpositum.
10. Anterior Cornu of the Lateral Ventricle.
11. Middle or Descending Cornu.
12. Posterior Cornu.
13. Hippocampus Major.
14. Tænia Hippocampi.
15. Hippocampus minor.
16. Longitudinal Fissure of the Brain.

tuous vessels partly covering the thalami nervorum opticorum, and continued into the inferior cornu of the

lateral ventricles. The plexus of each side is connected
to its fellow by the VELUM INTERPOSITUM, a membrane
which passes under the fornix, and lies on the third
ventricle and corpora quadrigemina.

From this plexus the blood is received by the VENA
GALENI situated in the middle of the velum interpositum
which consists of two parallel branches; these run backward, unite, and enter the TORCULAR HEROPHILI through
the SINUS RECTUS.

This plexus should now be detached at its forepart,
and turned back; it will remain as a guide to the knife
in tracing the inferior cornu of the lateral ventricle.

We now see

The CORPORA STRIATA, two smooth convexities, in
the forepart of the lateral ventricle, broad, and rounded
anteriorly, becoming narrow, and diverging as they pass
backward, consisting of tubular and vesicular substance
disposed in striæ.

The THALAMI NERVORUM OPTICORUM, two large oval
whitish eminences, placed by the side of each other between the diverging extremities or crura of the corpora
striata; toward their forepart is a peculiar eminence or
convexity, called the Anterior Tubercle. On the outer
and posterior face two enlargements, the CORPUS GENI-
CULATUM EXTERNUM AND INTERNUM. The former connected by a ridge to the testes. The latter to the nates.

TÆNIA SEMICIRCULARIS, a white medullary line,
running in the angle between the corpus striatum and
thalamus nervi optici of each side.

COMMISSURA ANTERIOR CEREBRI, a short cylindrical
medullary cord, stretched transversely between the fore
and lower part of the corpora striata, immediately under
the anterior crura of the fornix.

Just above the commissura anterior, and before the
thalami, is the VULVA, or foramen commune anterius, a
small slit or indentation, formed by the anterior crus of
the fornix, bifurcating, and inserting itself, on each side,
between the corpus striatum and thalamus nervi optici.
This slit is the space by which the three ventricles communicate.

3*

Commissura Mollis is an exceedingly soft, broad, cineritious junction between the convex surfaces of the thalami nervorum opticorum.

On separating the optic thalami, we discover the Third Ventricle. This is a longitudinal sulcus, or slit, situated between the thalami nervorum opticorum, and between the crura cerebri. Above, it is covered by the fornix and velum interpositum; at its upper and forepart, it communicates with the two lateral ventricles; below the commissura anterior, it opens into the infundibulum. This opening is termed Iter ad Infundibulum. Backward, it is continued by a canal which passes under the tubercula quadrigemina into the fourth ventricle. This passage is named Iter ad Quartum Ventriculum (aquæductus Sylvii).

The Pineal Gland, a small, soft, grayish, and conical body, of the size of a pea, is seated above the tubercula quadrigemina, and behind the thalami, to which it is connected by two white *penduculi*, or footstalks; its base is turned forward, and the apex backward; it is covered by the plexus choroides and posterior crura of the fornix. It contains an earthy matter, either in its own substance or that of the pedunculi, resembling sand. It was named by Sœmmering, who first discovered that it belongs to the healthy structure of the brain, the Acervulus Glandulæ Pinealis.

Commissura Posterior, a transverse cord at the back part of the third ventricle, before the tubercula quadrigemina, and above the iter ad quartum ventriculum.

Tubercula Quadrigemina, four small white bodies, adhering together, lying under the pineal gland, behind the third ventricle, and above the fourth. The uppermost two are named Nates, and the other two Testes.

From the under part of the testes, there project backward two ridges or cords, connecting themselves with the crura cerebelli, Processus a Cerebello ad Testes, and a thin medullary lamina between the Valve of Vieussens.

Fig. 10.

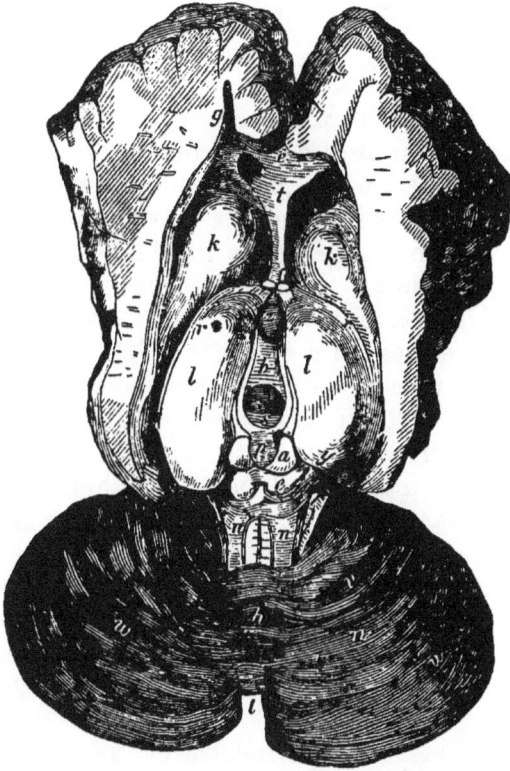

A Section of the Cerebrum, showing the Upper Surfaces of
the Corpora Striata and Optic Thalami, the Cavity of the
Third Ventricle, and the Upper Surfaces of the Cerebellum.

a, e. Tubercula Quadrigemina.	*s.* Posterior Commissure.
a. Nates.	*p.* Pineal Gland with its Pe-
e. Testes.	duncles.
b. Commissura Mollis.	*n, n.* Processus e Cerebello ad
c. Anterior Extremity of the	Testes.
Corpus Callosum cut.	*m, m.* Hemispheres of the Cere-
f. Crura of the Fornix.	bellum.
g. Anterior Horn of Lateral	*h.* Superior Vermiform Pro-
Ventricle.	cess.
k, k. Corpora Striata.	*i.* Notch between the Hemi-
l, l. Thalami Optici.	spheres of the Cerebel-
z to s. Third Ventricle.	lum behind.
In Front of *z* is the An-	
terior Commissure.	

The inferior cornu of the lateral ventricle, which descends into the middle lobe of the brain, may now be traced, by following the tract of the choroid plexus.[1] In it is seen

The HIPPOCAMPUS MAJOR, or Cornu Ammonis. At its commencement it is narrow, but it becomes a broad medullary projection of the floor of the ventricle; and its extremity, which is called PES HIPPOCAMPI, is curved inward. The thin edge on its inside, which follows the whole of its circuit, is named the CORPUS FIMBRIATUM, or Tænia Hippocampi. . Within this body is another, resembling the teeth of a comb, the CORPUS DENTATUM. The posterior crus of the fornix runs along its inner and anterior part, in the form of a thin floating edge.

In the posterior cornu of the lateral ventricle, which passes into the posterior lobe of the brain, there is a similar medullary projection, but smaller, the HIPPOCAMPUS MINOR.

Below the iter ad infundibulum are seen the CORPORA ALBICANTIA Willisii (corpora mammillaria), two medullary eminences of the size of peas. The remainder of these bodies is seen on the outer surface of the base of the brain.

This completes the demonstration of the cerebrum. The whole of the posterior lobes and the lateral part of the middle lobes may be removed. This exposes to your view

The TENTORIUM and the FALX CEREBELLI. At this point you should trace the bifurcation of the longitudinal sinus into the two lateral sinuses. The lateral sinuses are formed by the splitting of the laminæ of the tentorium; hence they follow the course of that membrane, run along their grooves in the occipital bone, and dip downward and forward through the foramen lacerum in basi cranii, to terminate in the internal jugular veins.

The TORCULAR HEROPHILI, or fourth sinus, runs along the middle of the tentorium, and joins the ex-

---

[1] Or it may be exhibited by cutting away successive slices of the side of the brain until the ventricle is exposed.

tremity of the longitudinal sinus at the point where it bifurcates.

The great notch of the tentorium is a circular opening left on the anterior part of the tentorium, allowing a junction between the cerebrum and cerebellum.

The brain should now be removed by elevating the front lobes, and dividing the nerves, arteries, etc., as they appear, cutting through the tentorium as it stretches along the Petrous bone, and, pushing the knife through the great occipital foramen, divide the spinal marrow and vertebral arteries, when it may be turned out and inverted. Clean it off well, leaving the nerves. Its under surface exhibits the ANTERIOR and MIDDLE lobes of the cerebrum, the two hemispheres of the cerebellum, the PONS VAROLII, or TUBER ANNULARE, and MEDULLA OBLONGATA.

CEREBELLUM.—This part of the brain, divided into two lobes by the *falx cerebelli*, or septum occipitale, is covered by a vascular membrane; consists of medullary and cineritious substance; but, instead of convolutions, has numerous deep sulci, into which the pia mater dips, and forms thin flat strata.

Remark the following processes:

1. Appendix, or PROCESSUS VERMIFORMIS SUPERIOR, situated under the pia mater, on the anterior and superior part of the cerebellum, the anterior part the MONTICULUS.

2. Appendix, or PROCESSUS VERMIFORMIS INFERIOR, will be found situated between the two lobes on the under surface of the cerebellum, and immediately behind the medulla oblongata.

On separating the two lobes behind, and making a deep incision, we discover

The FOURTH VENTRICLE.—The sides of this ventricle are formed by the cerebellum, the anterior part by the medulla oblongata, the upper and back part by the valvula cerebri; it is lined by a thin vascular membrane, and has on its forepart a groove or fissure, which, terminating in a sharp point, is named CALAMUS SCRIPTORIUS. On each side of this groove are seen several medullary lines, which are the origin of the portio mollis of the seventh pair of nerves. The iter a tertio ad

Fig. 11.

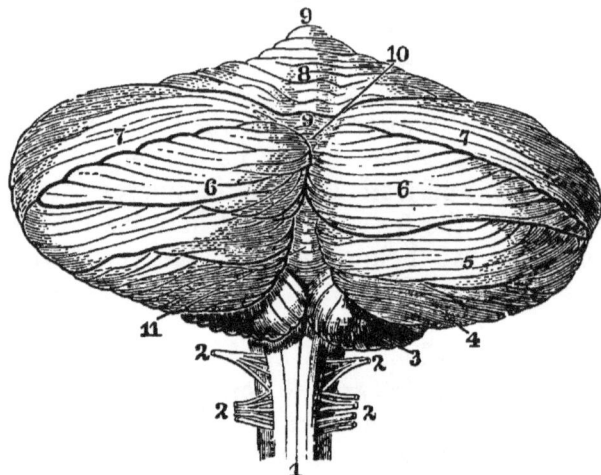

A POSTERIOR VIEW OF THE CEREBELLUM, AND THE SEVERAL LOBULES OF WHICH IT IS COMPOSED. (After Solly.)

1. Spinal Cord.
2. Posterior Spinal Nerves.
3. Amygdaloid Lobule.
4. Lobulus Pneumogastrici.
5. Lobulus Gracilis.
6. Inferior Semilunar Lobe.
7. Superior Semilunar Lobule.
8. Lobulus Quadratus.
9. Superior Vermiform Process.
10. Vermis Inferior.
11. Monticulus.

quartum ventriculum enters the upper part of the fourth ventricle. The valve of Vieussens hangs over it.

On cutting the cerebellum perpendicularly, there is formed, from the intermixture of cineritious and medullary matter, a tree-like appearance, named ARBOR VITÆ, of which the trunk is termed the peduncle of the cerebellum, and is continued to the back part of the medulla oblongata.

On the under surface is a gutter into which projects the INFERIOR VERMIFORM PROCESS—the VALLECULA. This leads to the fourth ventricle, on each side of which and in the centre are lobes. The lateral ones called the AMYGDALÆ, the intermediate one the UVULA. Another on the restiform body, the FLOCCULUS.

FISSURES OF SYLVIUS—Separate the anterior from the middle lobe of the cerebrum.

Fig. 12.
ANTERIOR ASPECT OF THE MEDULLA OBLONGATA.

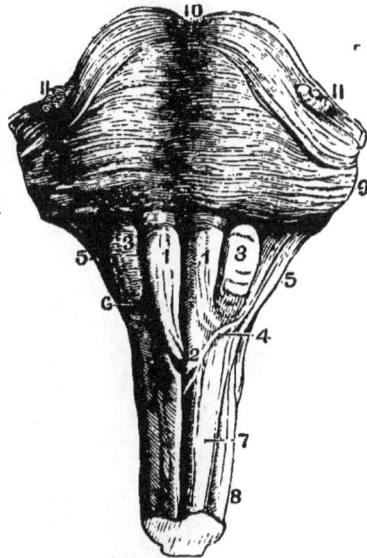

1. Corpora Pyramidalia.
2. The point of their Decussation.
3. Corpora Olivaria.
4. Fibres that run from the Anterior Column of the Medulla Spinalis to the Cerebellum.
5. Corpora Restiformia.
6. Arciform Fibres.
7. Anterior Columns.
8. Lateral Columns.
9, 10. Pons Varolii.
11. Roots of the Trigeminus Nerve.

MEDULLA OBLONGATA—Upper end of the Medulla Spinalis. On it a median fissure; three bodies on each side. One next to the fissure CORPUS PYRAMIDALE; next the OLIVARIS, next the RESTIFORME.

PONS VAROLII—The body at the top of the Medulla Oblongata. BASILAR ARTERY rests upon its middle.

CRURA CEREBELLI—Thick cords from the Pons to the cerebellum.

CRURA CEREBRI—Round cords passing in a divergent manner, from the Pons to the cerebrum.

CORPORA MAMMILLARIA or ALBICANTIA—Two little round bodies between the crura, front of the Pons.

LOCUS PERFORATUS—Perforated space between the last named bodies and the Pons.

TUBER CINEREUM or PONS TARINI—Triangular space in front of the corpora albicantia: a little process in its centre, the INFUNDIBULUM. In front the commissure of the optic nerves.

Fig. 13.

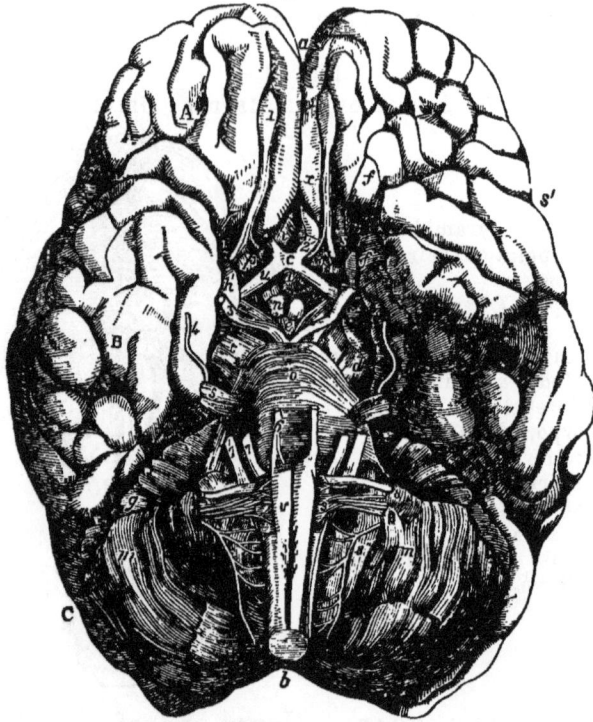

BASE OF THE BRAIN. A, ANTERIOR, B, MIDDLE, AND C, POSTERIOR
LOBES OF THE CEREBRUM.

a. Forepart of the Longitudinal or Inter-hemispheric Fissure.

b. Notch between the Hemispheres of the Cerebellum.

c. Optic Commissure.

d. Left Crus Cerebri.

e. Lobus Perforatus Lateralis.

e to *i.* Inter-crural Lamina.

ff. Convolution of the Fissure of Sylvius.

i. Infundibulum.

l. Right Crus Cerebelli.

m, m. Hemispheres of the Cerebellum.

n. Eminentia Mammillaria.

o. Pons Varolii, forming by its continuation on each side the Crus Cerebellum.

p. Pons Tarini.

q. Horizontal Fissure of the Cerebellum.

r. Gray Tuber.

s, s. Fissure of Sylvius.

t. Left Crus of the Cerebrum.

u, u. Optic Tracts.

v. Medulla Oblongata.

x. Marginal Convolution of the Longitudinal Fissure.

1. Olfactory Nerve.
2. Optic.
3. Motor Oculi.
4. Trochlearis or Patheticus.
5. Trigeminal or Trifacial.
6. Motor Externus.
7. Facial.
7. Auditory.
8. Glosso-pharyngeal.
8. Pneumogastric.
8. Spinal Accessory.
9. Hypoglossal,

SUBSTANTIA PERFORATA—Perforated space at commencement of the fissure of Sylvius.

Next examine the nerves.

1. The FIRST PAIR OF NERVES, the OLFACTORY, arise from the outside of the corpora striata, between the anterior and middle lobe of the brain; run under the anterior lobes, being lodged in two superficial grooves, and lying between the pia and dura mater; expand into a small oval ganglion, from which several small filaments descend through the cribriform plate of the ethmoid bone, to ramify on the membrane lining the nose.

2. The SECOND PAIR, the OPTIC, arise from the posterior part of the optic thalami, and also from the tubercula quadrigemina; they make a circle round the crura cerebri called the TRACTUS OPTICUS. The two nerves approach gradually, and unite, just before the pituitary gland, on the forepart of the sella turcica. They then diverge, and each nerve passes out at the foramen opticum of the sphenoid bone, to form the retina of the eye.

On each side of these nerves are seen the CAROTID ARTERIES. Each artery emerges from the cavernous sinus by the side of the anterior clinoid process: sends a branch forward, which, uniting with a similar branch of the other carotid, forms the anterior part of the CIRCULUS ARTERIOSUS WILLISII; while other branches, passing backward, and uniting with branches of the basilar artery, complete the posterior part of the arterial circle.

A fold of dura mater passes from the anterior to the posterior clinoid process of each side. This fold is double, and forms by its duplicature the CAVERNOUS SINUS.

On dividing the optic nerves, and inverting them, we see the infundibulum, a funnel of cineritious substance, leading from the inferior and anterior extremity of the third ventricle to the pituitary gland; it is generally imperforate before it reaches the gland.

The PITUITARY GLAND, a reddish body somewhat globular, consisting of two lobes, is situated in the sella turcica of the sphenoid bone, partly covered by a fold of dura mater, and attached to the infundibulum. The circular sinus is situated at this point.

3. The THIRD PAIR OF NERVES, MOTORES OCULORUM, arise from the crura cerebri,[1] pass outward and forward on the outer side of the posterior clinoid process into the cavernous sinus, and running through the foramen lacerum orbitale superius of sphenoid bone to the muscles of the eye.

Between these two nerves are seen the two vertebral arteries, ascending and uniting, to form the basilary artery.

4. The FOURTH PAIR, TROCHLEARES or PATHETICI, are very slender, and situated immediately under the edge of the tentorium. This nerve arises from the valve of Vieussens, comes out from betwixt the cerebrum and cerebellum, passes by the side of the pons Varolii, and passing through the cavernous sinus, continues its course through the foramen lacerum orbitale superius, to supply the obliquus superior muscle of the eye.

5. The FIFTH PAIR, TRIGEMINI, are much larger than the fourth, and are situated more outward and backward. Each of these nerves arises by a number of filaments, from the anterior and lowest part of the crus cerebelli, where the crus unites with the pons Varolii; it passes forward, enters the cavernous sinus, where it untwists itself, and forms a flat irregular ganglion, the GANGLION GASSERIANUM, and then divides into three great branches.

(1) RAMUS OCULARIS, or the OPHTHALMIC NERVE of Willis, passes through the foramen lacerum orbitale superius to the appendages of the eye.

(2) RAMUS MAXILLARIS SUPERIOR passes through the foramen rotundum to the upper jaw and face.

(3) RAMUS MAXILLARIS INFERIOR passes through the foramen ovale to the lower jaw and tongue.

6. The SIXTH PAIR, MOTORES OCULORUM EXTERNI, or abductores.—This nerve is small, but not so small as

---

[1] The two crura pass obliquely backward and inward, so as to converge and meet in front of the tuber annulare; it is from the hollow formed by their convergence, and named by Vicq d'Azyr, *fosse des nerfs oculo-musculaires*, that the third pair arise.

the fourth pair; it is seen arising betwixt the pons Varolii and corpora pyramidalia. It enters the cavernous
sinus; it there runs by the side of the carotid artery,
and passes through the foramen lacerum orbitale superius
to the rectus externus oculi.

7. The SEVENTH PAIR, NERVI AUDITORII, consist
of two portions.[1]

(1) The Portio Dura, or the facial nerve, arises from
the crus cerebelli, and comes out from the fossa or
groove betwixt the pons Varolii, corpora olivaria, and
crura cerebelli.

(2) The Portio Mollis, or more properly the auditory
nerve, arises from the inner surface of the fourth ventricle. It has a groove on its surface for receiving the
portio dura; accompanied by an artery, they enter the
meatus auditorius internus, where the portio mollis is
distributed to the parts of the internal ear, while the
portio dura runs through the aqueduct of Fallopius, and
comes out at the stylo-mastoid foramen below the ear, to
form the principal nerve of the face.

8. The EIGHTH PAIR consists of three, the Par Vagum,
Spinal Accessory, and Glosso-Pharyngeal. The PAR
VAGUM arises by numerous filaments from the sides of
the corpora olivaria and medulla oblongata. They unite,
run toward the foramen lacerum in basi cranii, pierce
the dura mater, and pass out through the anterior part
of the hole, having been first joined by the

NERVUS ACCESSORIUS, which runs up from the medulla spinalis through the great occipital foramen, and
then separate to their different destinations.

The great LATERAL SINUS passes out by the back
part of the same foramen, to form the internal jugular
vein; it is separated from the nerve by a slip of cartilage.

9. The NINTH PAIR, LINGUALES or HYPOGLOSSAL.—
This nerve arises from the furrow betwixt the corpora

---

[1] The classification of Sœmmering makes twelve pairs of cranial
nerves. Thus the Portio Dura and Mollis would be 7th and 8th
pairs. The Glosso-Pharyngeal the 9th, Pneumogastric the 10th.
The Spinal Accessory the 11th, and Hypoglossal the 12th.

olivaria and pyramidalia, by several filaments, which often pierce the dura mater separately. It passes through the anterior condyloid hole of the occipital bone to supply the muscles of the tongue.

Immediately after leaving the cranium, the eighth and ninth pair and the ganglion of the sympathetic are connected together.

Vessels of the Brain.

Internal carotid enters the skull through the carotid foramen in the temporal bone; passing through the cavernous sinus it divides into the following branches:

1. ANTERIOR CEREBRI, passing between the anterior cerebral lobes.

2. MEDIA CEREBRI, in the fissure of Sylvius.

3. ARTERIA COMMUNICANS, passing back to unite with the BASILAR.

VERTEBRAL ARTERY—From the subclavian ascends through the foramina of the cervical transverse processes, enters the foramen magnum, and gives branches to the dura mater and medulla spinalis, then

1. INFERIOR CEREBELLAR to under part of cerebellum. The union of the two vertebral form the

2. BASILAR, which rests on the pons Varolii.

3. SUPERIOR CEREBELLAR to upper part of cerebellum.

4. POSTERIOR CEREBRAL to the posterior lobes of the cerebrum. With this usually the communicating branch from the carotid unites to form the circle of Willis.

The veins open into the sinuses, and they into the *internal jugular vein.*

MEDULLA SPINALIS, or the Spinal Marrow. This part of the nervous or sensorial system must be here described, although its dissection cannot be performed till all the muscles of the back are removed, so that the posterior part of the spinal canal may be sawed off.

The spinal canal is lined by a strong ligamentous sheath, and the dura mater is continued down upon this sheath in the form of a funnel.

The spinal marrow consists externally of medullary substance, internally of vesicular. It runs down to the first lumbar vertebra, where it terminates by numerous filaments, which form the CAUDA EQUINA. It is closely embraced by the pia mater, while the tunica arachnoides adheres to that membrane very loosely. During the whole of its passage, there is on each side a membranous connection between the pia and dura mater, by distinct slips, irregular and pointed, which connection is named LIGAMENTUM DENTICULATUM. This ligament separates the two roots of the spinal nerves. The arteries of the medulla may be seen running down on its anterior and posterior surfaces; they are branches of the vertebral

Fig. 14.

FISSURES OR SULCI OF THE SPINAL MARROW.

1. Anterior Longitudinal Fissure.
2. Posterior Longitudinal Fissure.
3. Antero-Lateral Fissure, for the Corresponding Roots of the Spinal Nerves.
4. Postero - Lateral Fissure, for the Posterior Roots of the Spinal Nerves.
5. Lateral Fissure.

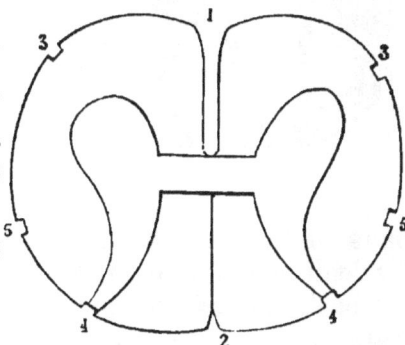

artery; and on these surfaces the ANTERIOR and POSTERIOR MEDIAN or LONGITUDINAL FISSURE, dividing it into lateral columns. Each column is again divided into three others by two lateral lines, the points where the roots of the spinal nerves come out. The cord has two enlargements upon it, the upper corresponding to the origin of the brachial plexus, the lower to the lumbar nerves.

The spinal accessory nerve is seen arising by small twigs from the posterior bundles of the fourth, fifth, sixth, and seventh cervical nerves; it then ascends along the spinal canal, enters the foramen, and passes forward, to accompany the par vagum.

The spinal marrow sends off thirty-one pairs of spinal nerves, which pass through the foramina formed between the bodies of the vertebræ. They consist of eight cervical, twelve dorsal, five lumbar, and five or six sacral

Fig. 15.

ORIGIN OF THE SPINAL NERVES.

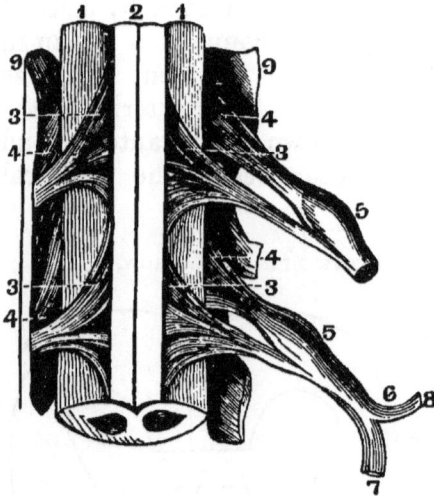

1, 1. Lateral Columns, marked off in front at (2) the Anterior Fissure
3. Anterior Roots.
4. Posterior Roots.
5. Ganglion formed by the Posterior Roots.
6. Spinal Nerve, formed by the junction of the Anterior and Posterior Roots.
7. Anterior Branch of the Spinal Nerve.
8. Posterior Branch.

pairs of nerves. Each of these nerves arises in two fasciculi or roots, one from the forepart, the other from the back part of the spinal cord. These fasciculi penetrate the dura mater separately. On the posterior bundle or root is a ganglion.

CHAPTER V.

DISSECTION OF THE ANTERIOR PART OF THE NECK.

Of the Muscles.

THE utility of this dissection must be evident, when you consider how many important parts are contained in the forepart of the neck. The tube which conveys air to the lungs, the vessels which are sent from the heart to the brain, and the nerves which are destined to supply the thoracic and abdominal viscera, are situated in the neck; and all these parts lie imbedded in cellular substance; hence the dissection is intricate, and requires the utmost care in its performance.

The muscles of the anterior part of the neck are sixteen in number on each side. They may be divided into muscles situated superficially, muscles at the upper part of the neck, and those situated at the lower part. To dissect the neck, make one incision from the top of the sternum to the symphysis of the chin, a second along the base of the jaw to the mastoid portion of the temporal bone, and a third from the sternum along the clavicle to the acromion process of the scapula. Reflect the integuments.

The superficial muscles are two.

Immediately under the integuments, and adhering to them.

1. The MUSCULUS CUTANEUS, vulgo, Platysma myoides—Will be found to be between two layers of the superficial fascia of the neck; the fascia extends over face and parotid gland. It *arises*, by slender separate fleshy fibres, from the cellular substance, covering the upper part of the deltoid and pectoral muscles. These fibres form a thin broad muscle, which runs obliquely upward, and is

Inserted into the skin and muscles covering the lower jaw and cheek. This muscle should be dissected in the course of its fibres; the skin, therefore, must be dissected off in an oblique direction from the clavicle to the chin.

Use. To draw the skin of the cheek downward, and, when the mouth is shut, to draw the skin under the lower jaw upward.

Remove the platysma myoides from its origin, and invert it over the face. Immediately beneath it is seen the external jugular vein, which is formed of branches from the temple, side of the face, and throat. It crosses obliquely over the sterno-mastoideus, passes behind the outer edge of that muscle, and plunges beneath the clavicle, to enter the subclavian vein.

If the platysma has been carefully lifted, several nerves may be exhibited in its deep surface. One toward the angle of the jaw.

SUPERFICIALIS COLLI—One to the ear just behind the jugular vein, the AURICULARIS MAGNUS; one along the posterior border of the S. cleido-mastoid muscle to the back of the head, the OCCIPITALIS MINOR; other branches descend; some of which are deep for the supply of muscles. All these are from the SUPERFICIAL CERVICAL PLEXUS, which comes out just behind the middle of the sterno-cleido-mastoid muscle, and is formed by the anterior branches of the three or four upper spinal nerves.

The strong fascia seen covering the muscles of the neck after the removal of the PLATYSMA. It forms sheaths for the muscles of the neck by detaching from its deep surface septa which pass in between them. It is attached behind to the spinous processes of the cervical vertebræ under the trapezius muscle; above to the base of the jaw as it passes to the face, and at its angle dips down to the styloid process, forming the STYLO-MAXILLARY LIGAMENT, and forming, as will be seen again, a complete separation between the PAROTID and SUBMAXILLARY GLANDS. Along the middle line of the neck it meets the fascia of the opposite side, incasing

muscles and thyroid gland; below it is attached both
to the top and inner surface of the sternum, to the
sternal end of the clavicle, to the cartilage of the first
rib, and is connected to the sheaths of the bloodvessels
and nerves as they pass into the axilla.

Fig. 16.

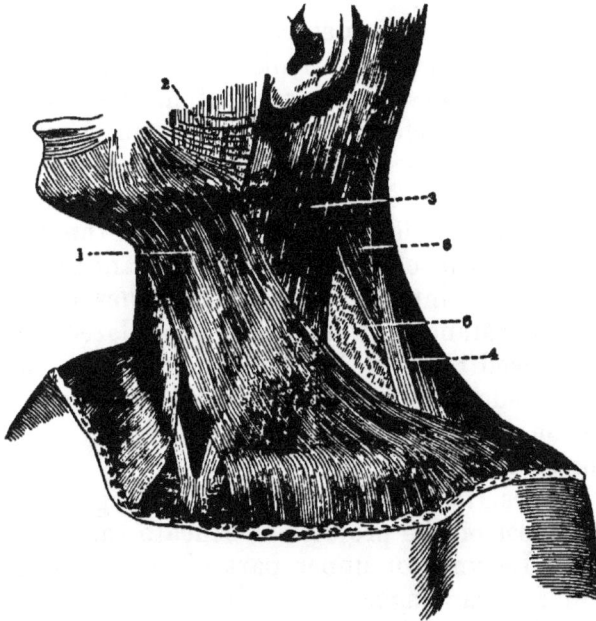

1. Platysma Myoides Muscle.
2. Scattered Fibres of the same, forming the Musculus Risorius of
   Santorini.
3. Sterno-Cleido-Mastoid Muscle.
4. Trapezius.
5. Splenius.
6. Posterior Large Triangle of the Neck, bounded in front by the
   Sterno-Cleido-Mastoid, behind by the Trapezius, and below by
   the Clavicle.

2. The STERNO-CLEIDO-MASTOIDEUS—*Arises* by two
distinct origins; the anterior, tendinous and fleshy, and
somewhat round, from the top of the sternum, near its
junction with the clavicle; the posterior or outer, fleshy
and flat, from the upper and anterior part of the clavicle. These two origins soon unite, and form a strong

4

muscle, which ascends obliquely upward and outward, to be

Inserted, tendinous, into the outside of the mastoid process, and into the transverse ridge behind that process.

Use. When one acts singly, it turns the head to one side. When both act together, they bend the head forward.

The muscle should be detached from the sternum and clavicle, and left suspended by its insertion. It is pierced by several branches of the cervical nerves, and, about its middle, it is perforated by the Nervus Accessorius. These nerves ramify on the neighboring muscles of the neck and shoulder. Between the posterior edge of the sterno-cleido-mastoideus and the forepart of the trapezius muscles, above the clavicle, is seen a quantity of loose fatty substance, intermixed with branches of nerves. This fatty substance is watery and granulated; it must not be removed roughly, lest important nerves and vessels be injured; it is continued around the vessels under the clavicle.

In the middle of the throat you see

(1) The Os Hyoides, or bone of the tongue, forming the uppermost of the projections beneath the chin.

(2) The Larynx, or upper part of the trachea, consisting of five cartilages, of which two are evident externally, viz.: 1. The uppermost and largest is the thyroid cartilage; 2. The inferior is the cricoid cartilage. The two arytenoid cartilages, and the epiglottis, lie behind this.

(3) The Trachea, consisting of cartilaginous rings, and extending into the thorax.

(4) Behind the larynx is situated the pharynx. At the part where the larynx terminates in the trachea, the pharynx contracts itself, and forms the œsophagus, or muscular tube, conveying the food to the stomach, which descends behind the trachea, situated rather to the left side of the cervical vertebræ.

These parts are covered by muscles, and on each side of the trachea lie the great vessels and nerves.

Muscles at the lower part of the neck are five.

3. The STERNO-HYOIDEUS — *Arises*, thin and fleshy, from the upper and inner part of the sternum, clavicle, and first rib; it forms a flat and narrow muscle.

Inserted into the base of the os hyoides.

Situation. This pair of muscles is seen on removing the platysma myoides, between the sterno-cleido-mastoidei.

Use. To pull the os hyoides downward.

4. The OMO-HYOIDEUS, inclosed in lamina of the deep fascia—*Arises*, broad, thin, and fleshy, from the root of the coracoid process, and semilunar notch of the scapula, ascends across the neck, and forms a middle tendon, where it passes below the sterno-cleido-mastoideus. Becoming fleshing again, it runs up, and is

Inserted into the base of the os hyoides, between its cornu and the insertion of the sterno-hyoideus.

Situation. The lower part of this muscle is covered by the trapezius; its middle by the sterno-cleido-mastoideus; its anterior part is seen on removing the platysma myoides; it crosses over the carotid artery and internal jugular vein.

Use. To pull the os hyoides obliquely downward.

5. The STERNO-THYROIDEUS. — This muscle *arises*, fleshy, from the inside of the sternum, and of the extremity of the first rib; forms a flat muscle, and is

Inserted into the inferior edge of the oblique ridge on the ala or side of the thyroid cartilage.

Situation. Beneath the sterno-hyoideus.

Use. To draw the thyroid cartilage, and consequently the larynx, downward.

Under the sterno-thyroideus we find situated the

THYROID GLAND, a large reddish mass, situated on the superior rings of the trachea, below the cricoid cartilage; in form somewhat like a crescent, with the cornua turned upward.

6. The THYRO-HYOIDEUS—*Arises*, fleshy, from the upper surface of the oblique ridge in the ala of the thyroid cartilage, and passes upward, to be

Inserted into part of the base, and almost all the cornu of the os hyoides.

Use. To pull the os hyoides downward, or the thyroid cartilage upward.

Situation. Concealed by the sterno-hyoideus.

7. The CRICO-THYROIDEUS—*Arises*, tendinous and fleshy, from the side and forepart of the cricoid cartilage, and runs obliquely upward.

Inserted, by two fleshy portions, the first into the lower part of the thyroid cartilage, and the second into its inferior cornu.

Situation. On the side of the larynx, and under the sterno-thyroideus.

Use. To pull forward and depress the thyroid, or to elevate and draw backward the cricoid cartilage.

The muscles at the upper part of the neck consist of nine pairs.

8. The DIGASTRICUS—*Arises*, principally fleshy, from the fossa at the root of the mastoid process of the temporal bone; its fleshy belly terminates in a strong round tendon, which runs downward and forward, passes through the fleshy belly of the stylo-hyoideus, is fixed by strong ligamentous and tendinous fibres to the os hyoideus, from which it receives an addition of fibres; it then becomes again fleshy, and runs obliquely upward, to be

Inserted into a rough sinuosity on the anterior inferior edge of that part of the lower jaw called the Chin.

Situation. Its posterior belly is covered by the sterno-cleido-mastoideus; its anterior belly lies immediately under the skin and platysma myoides.

Use. To open the mouth, by pulling the lower jaw downward and backward; and, when the jaws are shut, to raise the larynx, and consequently the pharynx upward, as in deglutition.

In the triangular space formed by the two bellies of this muscle, and the base of the lower jaw, lies the SUB-MAXILLARY GLAND. It lies upon a flat muscle, the mylo-hyoideus, which is seen between the two bellies of the digastricus; the gland is surrounded by little absorbent glands, and is incased in a pocket of deep fascia.

9. The STYLO-HYOIDEUS—*Arises*, tendinous, from

Fig. 17.

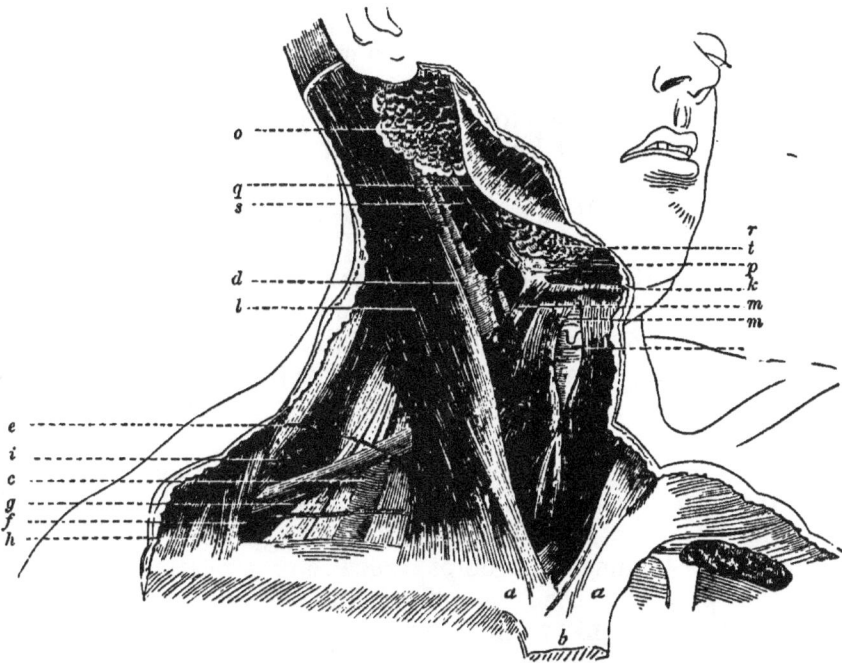

LATERAL VIEW OF THE NECK, WITH THE SKIN, PLATYSMA, AND CER-
VICAL FASCIA REMOVED.

a, a. Sternal ends of the Clavicles.
b. Upper part of the Sternum.
c. Third portion of the Subclavian Artery.
d. Primitive Carotid Artery dividing into the internal and
   external Carotids.
e. Posterior Scapular Artery.
f. Transverse Cervical or Supra-scapular Artery.
g. Brachial Plexus of Nerves.
h. Trapezius Muscle.
i. Tendon of the Omo-hyoid
   Muscle.

k. Anterior belly of the Omo-
   hyoid.
l. Sterno-Cleido-Mastoid Muscle.
m, m. Sterno-hyoid Muscles.
n. Larynx.
o. Parotid Gland.
p. Submaxillary Gland.
q. Posterior Belly of the Digastric Muscle.
r. Anterior Belly of the same.
s. Stylo-hyoid Muscle.
t. Hyoid Bone.

the middle and inferior part of the styloid process of
the temporal bone; its fleshy belly is generally perforated by the digastricus.

Inserted, tendinous, into the os hyoideus at the juncture of its base and cornu.

Situation. The origin of this muscle is situated more inward than the last; it is the most superficial of three muscles which arise from the styloid process; sometimes it is accompanied by another small muscle, having the same origin and insertion, THE STYLO-HYOIDES ALTER.

Use. To pull the os hyoides to one side, and a little upward.

10. The STYLO-GLOSSUS—*Arises*, tendinous and fleshy, from the styloid process, and from a ligament that connects that process to the angle of the lower jaw. It descends, and becomes broader, but less thick.

Inserted into the root of the tongue, runs along its side, and is insensibly lost near its tip.

Situation. This muscle lies within and rather above the stylo-hyoideus; underneath it is a ligament, extending from the styloid process to the angle of the os hyoides. Ligamentous fibres are also seen passing from that process to the inside of the angle of the lower jaw, STYLO-MAXILLARY LIGAMENT.

Use. To move the tongue laterally and backward.

11. The STYLO-PHARYNGEUS—*Arises*, fleshy, from the root and inner part of the styloid process.

Inserted into the side of the pharynx and back part of the thyroid cartilage.

Situation. It is situated deeper, and behind the styloglossus.

Use. To dilate and raise the pharynx and thyroid cartilage upward.

On removing the submaxillary glands, and detaching the digastric muscle from the os hyoides and chin, we expose the next muscle; but should notice in its removal the facial artery which runs through it on its way to the face, giving to it the SUBMAXILLARY BRANCHES, and one which runs along the jaw toward the chin, the SUBMENTAL.

12. The MYLO-HYOIDEUS—*Arises*, fleshy, from all the inside of the lower jaw, between the last molar tooth and the middle of the chin; the fibres form a flat muscle, converge, and are

Inserted into the lower edge of the base of the os hyoides.

Situation. This muscle unites with its fellow in a middle tendinous line which extends from the os hyoides to the chin; its posterior part is lined by the internal membrane of the mouth; it lies under the digastricus, but is seen betwixt its bellies.

Use. To pull the os hyoides forward, upward, and to either side.

The submaxillary gland sends off a duct, which passes behind the posterior edge of the mylo-hyoideus, and runs along the inner surface of this muscle forward and upward, on the inside of the sublingual gland, to open into the mouth on the side of the frenum of the tongue. Above it will be seen the LINGUAL BRANCH of the inferior maxillary nerve.

The SUBLINGUAL GLAND lies immediately above the mylo-hyoideus, betwixt it and the internal membrane of the mouth, where it lines the side and inferior surface of the tongue. It sends off several ducts, which open into the mouth, between the root of the tongue and side of the lower jaw.

The removal of the mylo-hyoideus exposes a pair of muscles, which are closely attached to one another.

13. The GENIO-HYOIDEUS—*Arises*, tendinous, from a projection on the inside of that part of the lower jaw which is called the Chin; it descends, becoming broader, and is

Inserted into the basis of the os hyoides.

Use. To draw the os hyoides forward and upward to the chin.

By removing this muscle, or turning it back from its origin from the jaw, we discover the next muscle.

14. The GENIO-HYO-GLOSSUS — *Arises*, tendinous, from a rough protuberance on the inside of the lower jaw, higher up than the origin of the genio-hyoideus; its fibres run forward, upward, and backward, in a very wide and radiated manner, to be

Inserted, some into the posterior part of the base of the os hyoides, near its cornu, others into the tip, middle, and root of the tongue.

Situation. This muscle lies under the genio-hyoideus before, and more outwardly under the mylo-hyoideus.

Fig. 18.

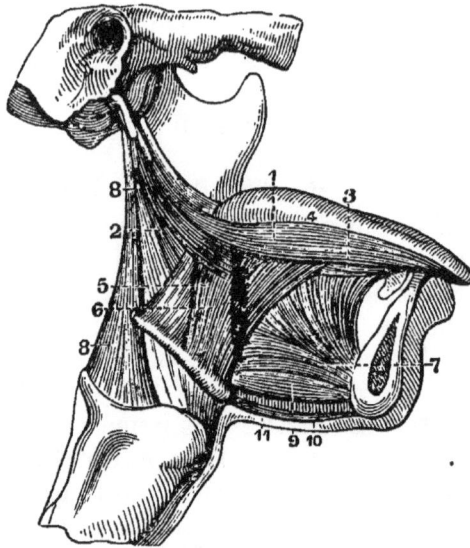

MUSCLES OF THE TONGUE.

1. Stylo-glossus.	7. Genio-hyo-glossus.
2. Stylo-hyoideus.	8. Stylo-pharyngeus.
3. Lingualis.	9. Genio-hyoideus.
4. Dorsum of the Tongue.	10. Raphé of the Mylo-hyoideus.
5, 6. Hyo-glossus.	11. Digastricus.

Use. According to the direction of its fibres, to draw the tip of the tongue backward into the mouth, the middle downward, and to render its dorsum concave; to draw its root and the os hyoides forward, and to thrust the tongue out of the mouth.

15. The HYO-GLOSSUS — *Arises*, broad and fleshy, from half of the base, and part of the cornu of the os hyoides; the fibres run upward, to be

Inserted into the inside of the tongue. Be careful here of the Hypoglossal Nerve, and Lingual Artery, to be presently described.

Situation. It is situated more outwardly than the

genio-hyo-glossus, and, at its insertion into the tongue, mixes with the stylo-glossus.

Use. To move the tongue inward and downward.

16. The LINGUALIS—*Arises*, from the root of the tongue laterally, and runs forward between the hyoglossus and genio-hyo-glossus, to be

Inserted into the tip of the tongue, along with part of the stylo-glossus.

Use. To contract the substance of the tongue, and bring it backward.

Of the Vessels and Nerves seen in the Dissection of the Neck.

Arteries.

The CAROTID ARTERY ascends from the thorax by the side of the trachea; on its outer side it has the internal jugular vein and the par vagum, and behind the sympathetic nerves. All these parts are connected and inclosed by condensed cellular membrane, which forms a kind of sheath for containing them. At the bottom, and in the middle of the neck, the carotid is covered by the sterno-cleido-mastoideus; at the upper part, by adipose membrane, absorbent glands, and by the platysma myoides. It lies deep on the muscles of the spine, and gives off no branches, until it reaches the space between the larynx and angle of the jaw, just below the cornu of the os hyoides, where it divides into the external and internal carotids.

Of the two, the internal carotid is situated most outwardly; it passes deep to the base of the cranium, where it enters the foramen caroticum, to supply the brain.

The external carotid immediately begins to send off branches.

Anteriorly it sends off

1. The A. THYROIDEA SUPERIOR.—This artery passes downward and forward, to ramify on the thyroid gland, where it inosculates with the artery of the other side, and with the inferior thyroid arteries. In this course, it

sends ramifications to the integuments, the outside of the larynx, the muscles, etc., and one remarkable branch, the A. LARYNGEA, which sometimes arises from the trunk of the external carotid; it is a small artery which divides betwixt the os hyoides and thyroides cartilages, to supply the internal parts of the larynx.

2. A. LINGUALIS.—This artery passes over the cornu of the os hyoides, then parallel with it, covered by the stylo-glossus, hyo-glossus, and genio-hyo-glossus, to which it gives branches, and terminates in running along the inferior part of the tongue from its base to its apex. It gives branches to the muscles about the chin, and to the substance and back part of the tongue.

3. A. FACIALIS. — The external maxillary artery passes under the stylo-hyoideus and posterior belly of the digastricus, then buries itself under the submaxillary gland. It runs over the lower jaw before the anterior edge of the masseter muscle, to supply the face.

In its passage, it gives off numerous branches. One is worthy of notice, which runs along under the line of the lower jaw,

The SUBMENTAL, and passing over the jaw near the symphysis, supplies the chin. Other twigs supply the submaxillary gland, etc.

Posteriorly the carotid sends off

4. A. OCCIPITALIS. — The occipital artery crosses backward and upward, over the internal jugular vein and internal carotid artery, under the belly of the digastric muscle; it passes through a slight groove in the mastoid process, below its great fossa, and ramifies on the back part of the head. A remarkable branch of the occipital passes toward the base of the skull, to inosculate near the foramen magnum occipitis, with branches from the vertebral and posterior cervical arteries.

5. A. PHARYNGEA is a small branch of the carotid, which passes inward to the pharynx and base of the skull.

6. POSTERIOR AURIS, which passes backward and upward in the fold, between the ear and scalp, and is distributed on the integuments of the head.

The external carotid ascends behind the angle of the jaw, and enters the parotid gland, where it divides into

7. A. MAXILLARIS INTERNA.
8. A. TEMPORALIS.
9. A. TRANSVERSALIS FACIEI.

These arteries are described in the dissection of the face.

Veins.

The INTERNAL JUGULAR VEIN is a continuation of the lateral sinus which passes through the foramen lacerum in the base of the cranium. It comes out deep from under the angle of the jaw, and in its course down the neck, it runs on the outer side of the carotid artery, before it reaches the thorax; it passes rather more forward than the artery, to join the subclavian vein.

Its branches accompany the ramifications of the external carotid. At first the vein which accompanies each artery is a single branch, but it soon subdivides, so that two veins accompany one artery.

Nerves.

1. The EIGHTH PAIR, or Par Vagum.—On separating the internal jugular vein, and trunk of the carotid artery, the par vagum is seen lying in the same sheath of cellular substance with those vessels. It lies in the triangular space formed betwixt the back part of the artery and vein, and the subjacent muscles. This nerve comes out of the foramen lacerum with the jugular vein; hence it adheres to that vein more closely than to the artery or muscles; it runs down the neck behind these vessels.

In this course it gives off several nerves.

(1) At the base of the cranium it sends off several filaments, which are connected with the other nerves coming out of the base of the skull, such as the ninth pair, the superior cervical ganglion of the sympathetic, etc.

(2) NERVUS GLOSSO-PHARYNGEUS leaves the other branches of the eighth pair, deep under the angle of the

jaw. It passes behind the carotids toward the muscles arising from the styloid process; one principal branch of it passes between the stylo-pharyngeus and stylo-glossus to the tongue, while other twigs run behind the stylo-pharyngeus, to supply the pharynx.

Fig. 19.

NERVES OF THE NECK AND TONGUE.

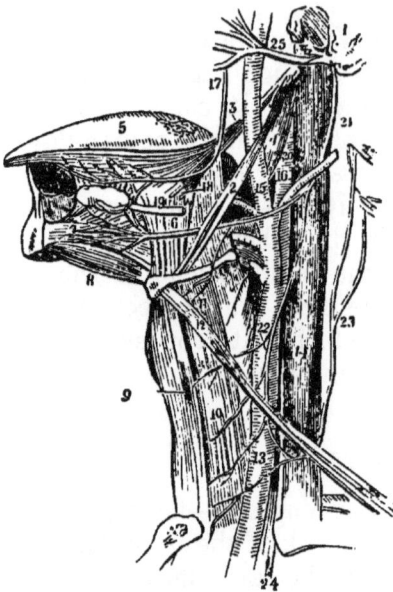

1. Part of the Temporal Bone.
2. Stylo-hyoid Muscle.
3. Stylo-glossus Muscle.
4. Stylo-pharyngeus Muscle.
5. Tongue.
6. Hyo-glossus Muscle.
7. Genio-hyo-glossus Muscle.
9. Sterno-hyoid Muscle.
10. Sterno-thyroid Muscle.
11. Thyro-hyoid Muscle, upon which is seen a branch of the Hypoglossal Nerve.
12. Omo-hyoid Muscle straightened by the removal of the Loop of Cervical Fascia through which its tendon plays.
13. Common Carotid Artery.
14. Internal Jugular Vein.
15. External Carotid Artery.
16. Internal Carotid.
17. Gustatory Branch of the Fifth Nerve, giving a Branch to (18) the submaxillary Ganglion.
19. Duct of Submaxillary Gland.
20. Glosso-pharyngeal Nerve.
21. Hypoglossal Nerve.
22. Descending Branch from the Cervical Plexus.
23. Communicating Branch from the Cervical Plexus.
24. Pneumogastric Nerve emerging from between the Internal Jugular Vein and Common Carotid Artery to enter the Chest.
25. Facial Nerve, emerging from the Stylo mastoid Foramen, and crossing the External Carotid Artery.

(3) NERVUS LARYNGEUS SUPERIOR.—The superior or internal laryngeal nerve passes behind the internal carotid artery, obliquely, downward and forward; then, under the hyo-thyroideus muscle, it plunges betwixt the os hyoides and thyroid cartilage, accompanying the laryn-

geal artery, and supplying the internal parts of the larynx.

(4) In the neck, also, the par vagum gives off filaments to the cervical ganglions of the sympathetic nerve, and communicates with the other nerves of the neck. Filaments also unite with twigs of the sympathetic, and run down over the carotid artery to the great vessels of the heart, where they form the superior cardiac plexus.

The par vagum enters the thorax by passing betwixt the subclavian artery and vein.

2. The GREAT SYMPATHETIC NERVE. — This nerve lies behind the carotid, in the cellular membrane, betwixt that vessel and the muscles covering the vertebræ of the neck.   It is distinguished from the par vagum by being smaller, lying nearer the trachea, and adhering to the muscles of the spine; also by its forming several ganglions.   It comes out by the same foramen as the carotid artery.

Immediately after its exit from the skull, it forms the SUPERIOR CERVICAL GANGLION, which is very long, and of a reddish color. The nerve afterward becomes smaller, and descends; and opposite the fifth or sixth cervical vertebra, it forms another swelling, the INFERIOR CERVICAL GANGLION.   Sometimes it has another ganglion about the fourth or fifth vertebra of the neck, the MIDDLE CERVICAL GANGLION: but this is not a constant appearance.   The nerve then passes behind the subclavian artery into the thorax.

The branches of the intercostal nerve are numerous, and they generally pass off from the ganglions.   Immediately below the base of the cranium, twigs go to the eighth and ninth pairs, and to the upper cervical nerves. In the middle of the neck, some twigs pass over the carotid; others go to the parts covering the trachea, uniting with filaments of the par vagum; others unite with the descendens noni, or descending branch of the ninth pair, and some filaments assist the twigs of the par vagum to form the superior cardiac nerve.   In the lower part of the neck, twigs are sent to communicate with the cervical nerves, etc.

Fig. 20.

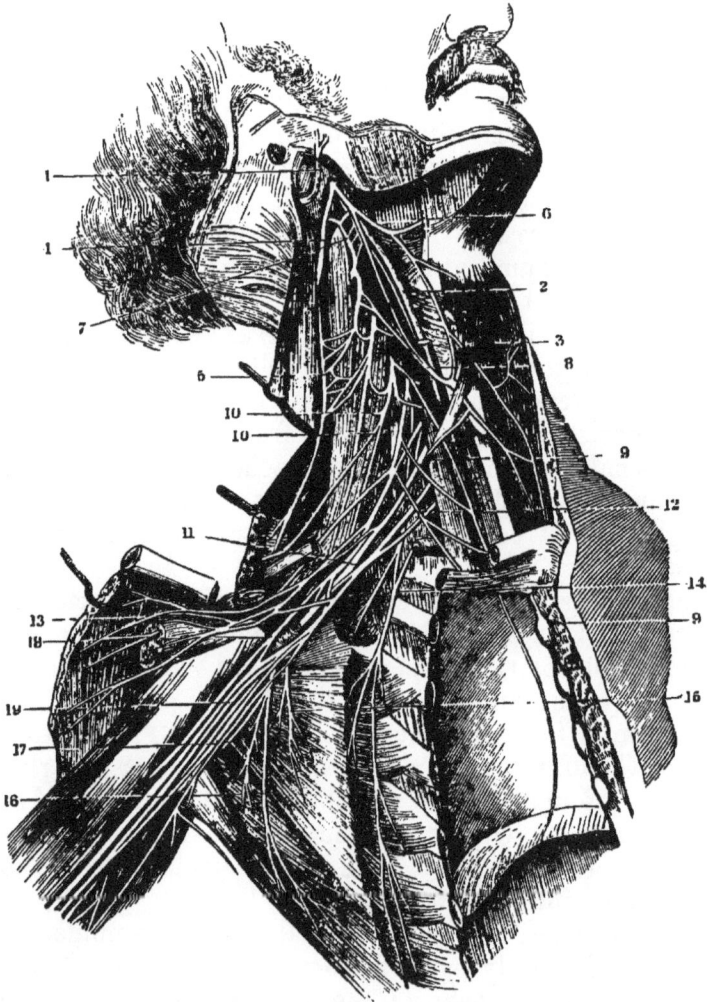

CERVICAL AND BRACHIAL PLEXUSES OF NERVES OF THE RIGHT SIDE.

1. Facial Nerve.
2. Pneumogastric Nerve.
3. Internal Carotid Artery.
4. Spinal Accessory Nerve.
5. Anastomoses of the Spinal Accessory Nerve with the Cervical Plexus.
6. Hypoglossal Nerve, giving off its Descending Branch.

7. Anterior Branch of the first Cervical Nerve, anastomosing with the Hypoglossal Nerve and with the Pneumogastric.
8. Descending Cervical Branch of the Cervical Plexus, anastomosing with the corresponding Branch of the Hypoglossal.
9. Phrenic Nerve.
10, 10. Deep Cervical Branches of the Cervical Plexus.
11. Brachial Plexus.
12. Branch to the Subclavian Muscle, sending a Filament to the Phrenic Nerve.
13. Anterior Thoracic Branches.
14. Lateral Thoracic Branch, or the Branch to the Great Serrate Muscle.
15, 16, 17. Subscapular Branches going to the Subscapular, Latissimus, and Greater Teres Muscles.
18. Axillary Artery, surrounded by a sort of Sheath, formed by Branches going to the Arm.
19. Brachial Branches.

3. The NINTH PAIR, Nervus Hypoglossus, or Lingual Nerve, having left the skull by the anterior condyloid foramen, is connected with the eighth pair and sympathetic nerve. Like them, it lies deep, and comes out from under the angle of the jaw. It is seen passing from behind the internal jugular vein, and then over the carotid artery, running betwixt these two vessels. It next passes under the mylo-hyoideus, running over the styloglossus, hyo-glossus, and genio-hyo-glossus, which last muscle its numerous branches perforate.

4. BRANCHES.—While the nerve is passing betwixt the jugular vein and the carotid artery, it sends off the DESCENDENS NONI.—This small and delicate nerve descends on the forepart of the vein and artery, and is distributed to the muscles on the anterior part of the trachea. It is joined by a filament formed by the first, second, and third cervical nerves, the COMMUNICANS NONI.

The LINGUAL BRANCH sent off by the third branch of the fifth pair of nerves, is also seen in the dissection of the neck. It is found under the mylo-hyoideus; it lies close upon the lower edge of the jaw-bone, betwixt the inferior edge of the pterygoideus internus and the upper part of the mylo-hyoideus. It gives numerous twigs to the sublingual gland and submaxillary duct, which are situated near it, and is lost in the substance of the tongue.

5. Nervus Accessorius ad par Vagum, one of the eighth pair.—The accessory nerve, having passed out of the cranium with the par vagum, separates from it, passes behind the internal jugular vein obliquely downward and backward; it perforates the mastoid muscle, and is distributed to the trapezius and muscles about the shoulder; it is much connected with the third and fourth cervical nerves.

6. The Seven Cervical Nerves come out from the foramina betwixt the vertebræ of the neck. They send numerous branches to the muscles, etc. on the side of the neck, and communicate by filaments with all the other nerves in the neck.

In this stage of the dissection we may also see

7. The Phrenic Nerve, formed by branches of the third and fourth cervical nerves. This small nerve lies upon the belly of the Scalenus Anticus Muscle, and passes into the thorax, betwixt the subclavian artery and vein.

8. Recurrent Nerve.—From the par vagum in the thorax, lying close alongside the trachea as it ascends.

9. Brachial Plexus.—Seen between the Scaleni Muscles, formed by four lower cervical, and first dorsal spinal nerves, passes into the axilla to become the *axillary plexus*.

Arteries.

Subclavian Artery on the right side is a branch of the Innominata, on the left it is derived from the Aorta; on the right side it passes by a gentle curve to the scalenus anticus muscle, and rests upon the pleura. This first part on the left side rises almost perpendicularly up out of the chest, and has the pleura in front. Both after this pass beneath the scalenus anticus, having the brachial plexus of nerves and the scalenus posticus to the outer and posterior side, thence over the first rib, after which they become Axillary Arteries.

Thyroid Axis—A stem which generally furnishes

1. Inferior Thyroid Artery, which passes under the carotid bloodvessels to the thyroid gland.

Fig. 21.

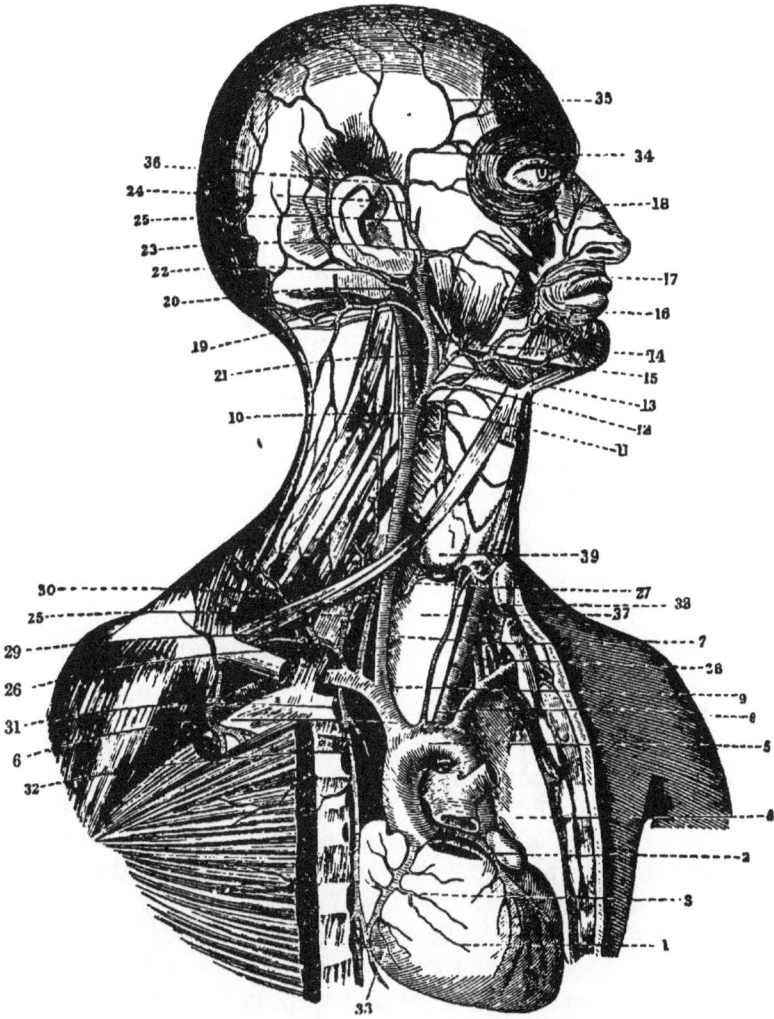

ARTERIES OF THE NECK.

1. Heart.
2. Left Coronary Artery.
3. Right Coronary Artery.
4. Pulmonary Artery cut through.
5. Arch of the Aorta.
6. Innominata Artery.
7. Right Primitive Carotid.
8. Left Subclavian.
9. Division of the Innominata into the Right Primitive Carotid and Right Subclavian.
10. Division of the Primitive Carotid into External and Internal Carotid.

11. Superior Thyroid Artery.
12. Lingual Artery.
13. Facial or External Maxillary Artery.
14. Inferior Palatine Artery.
15. Submental Artery.
16. Inferior Labial Artery.
17. Superior Labial Artery.
18. Lateral Nasal Branch.
19. Occipital Artery.
20. Posterior Auricular Artery.
21. Ascending Pharyngeal Artery.   34,
22. Division of the External Carotid into Temporal and Internal Maxillary Artery.
23. Transverse Facial Artery.
24. Temporal Artery.
25. Middle or Deep Temporal Artery.
25. Inferior Thyroid Artery.
26. Vertebral Artery.

27. Point at which the Vertebral Artery enters the opening in the Transverse Process of the Sixth Cervical Vertebra.
28. Left Superior Intercostal Artery.
29. Transverse Cervical Artery.
30. Posterior Scapular Artery.
31. Internal Mammary Artery.
32. Mediastinal Branch.
33. Superior Phrenic Artery.
35. Anterior Temporal Artery.
36. Posterior Temporal.
37. Trachea.
38. Middle Thyroid Artery, an anomalous branch of the aorta sometimes met with.
39. Thyroid Body.
40. Ascending Cervical Artery, a branch of the Inferior Thyroid.

2. SUPERFICIALIS CERVICIS—distributed to the deep layer of muscles front of spine.

3. SUPRA-SCAPULAR—following the course of the clavicle, it supplies the muscles on the dorsum of the scapula.

4. POSTERIOR SCAPULAR—across the neck, along the posterior border of the scapula—supplies the muscles on that portion of the scapula.

PROFUNDA CERVICIS—passes back between the transverse processes of last cervical and first dorsal vertebræ, it sends a branch up to inosculate with the PRINCEPS CERVICIS from the OCCIPITAL ARTERY.

Backward, the subclavian sends off

The VERTEBRAL ARTERY.—This artery arises from the back part of the subclavian, passes outward, and enters the foramen in the transverse process of the sixth cervical vertebra, and ascends through the tranverse processes of the vertebræ, to enter the foramen magnum of the occipital bone.

Anteriorly, the subclavian artery gives off

5. A. MAMMARIA INTERNA.—The internal mammary arises from the forepart of the subclavian opposite the

cartilage of the first rib; it runs down on the inside of
the cartilages of the ribs, and terminates in the abdominal muscle, where it inosculates with the epigastric.   It
is a large artery, and its branches are very numerous.
They pass to the external muscles of the chest, to the
intercostal muscles, pleura, etc.   It also sends off the
ARTERIA PHRENICA SUPERIOR, vel comes nervi phrenici,
which, with two veins, accompanies the phrenic nerve to
the diaphragm.

6. The subclavian artery gives twigs to the root of the
neck, and to the muscles about the scapula.

7. A. INTERCOSTALIS SUPERIOR.—Frequently a trunk
comes off from the subclavian, especially on the right
side, which passes downward and backward, and lodges
itself by the spine, to supply the two or three superior
intercostal spaces.

Course of the Subclavian Vein.

The SUBCLAVIAN VEIN is situated anteriorly to the
subclavian artery; it passes inward behind and under the
subclavius muscle, and before and over the belly of the
anterior scalenus (so that this last muscle lies betwixt the
vein and artery).   It runs over the first rib, where it is
found in contact with the axillary artery, and is called
the Axillary Vein.

The branches of this vein accompany and correspond
to the ramifications of the subclavian artery, returning
the blood from the thyroid gland, neck, chest, intercostal spaces, etc.   The subclavian vein also receives
the internal jugular, which passes down behind the
clavicle.

The Course of the Brachial Plexus of Nerves

May also be examined.   This plexus is formed by
branches of the four lower cervical and first dorsal
nerves, which pass between the anterior and middle
scaleni muscles into the axilla.   In this passage they
are situated higher up than the artery.

A considerable part of the scaleni muscles may now be seen, covered by a strong fascia, the PRÆVERTEBRAL FASCIA; the upper insertion of these muscles must be dissected with the muscles of the back part of the neck.

1. The SCALENUS ANTICUS—*Arises*, by three tendons, from the transverse processes of the fourth, fifth, and sixth vertebræ of the neck.

Inserted, tendinous and fleshy, into the upper edge of the first rib, near its cartilage.

2. The SCALENUS MEDIUS—*Arises*, tendinous, from the transverse processes of all the vertebræ of the neck.

Inserted into the upper and outer part of the first rib, from its root to within the distance of half an inch from the scalenus anticus.

3. The SCALENUS POSTICUS—*Arises*, tendinous, from the transverse processes of the first and sixth vertebræ of the neck.

Inserted into the upper edge of the first rib, near the spine.

Situation. These muscles are covered before by the sterno-mastoideus and trapezius, behind by the trapezius and levator scapulæ; but the scaleni are so connected with the muscles of the spine that the whole of them cannot be demonstrated till the muscles of the back and neck are dissected.

Uses of these three muscles: to bend the neck to one side, and, when the muscles of both sides act, to bend it forward; or, when the neck is fixed, to elevate the ribs, and dilate the chest.

Dissection of the Muscles on the front of the Cervical Vertebræ.

These muscles may be exposed by raising off the muscles already dissected, together with the trachea, larynx, and œsophagus. There are four pairs.

1. LONGUS COLLI—*Arises*, tendinous and fleshy, from the sides of the bodies of the three cervical vertebræ, and from the anterior surface of the transverse processes of the four or five lower cervical vertebræ.

Inserted, tendinous and fleshy, into the forepart of the bodies of all the vertebræ of the neck.

Situation. This muscle lies behind the œsophagus and great vessels and nerves of the neck.

Use. To bend the neck forward and to one side.

Fig. 22.

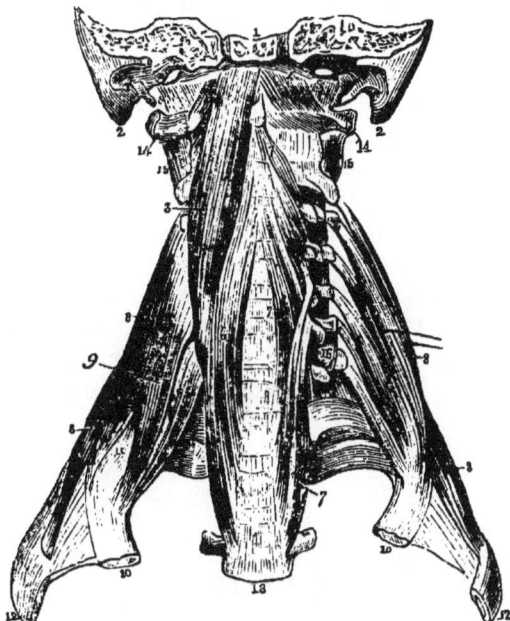

CERVICAL MUSCLES.

1. Basilar Process of the Occipital Bone.
2. Mastoid Process.
3. Rectus Capitis Anticus Major.
4. Rectus Capitis Anticus Minor.
5. Rectus Capitis Lateralis.
6. Longus Colli, right side.
7. Same, left side.
8, 8. Scalenus Posticus.
9. Scalenus Anticus.
10. First Rib.
11. Passage of the Subclavian Artery
12. Second Rib.
13. Third Dorsal Vertebra.
14. Transverse Process of the Atlas.
15. First Inter-transversalis Muscle.
16. Sixth Inter-transversalis Muscle.

2. RECTUS CAPITIS ANTICUS MAJOR—*Arises*, tendinous and fleshy, from the anterior points of the trans-

verse processes of the third, fourth, fifth, and sixth cervical vertebræ.

Inserted into the cuneiform process of the os occipitis in front of the condyloid process.

Situation. Farther out than the longus colli.

Use. To bend the head forward.

3. RECTUS CAPITIS ANTICUS MINOR—*Arises*, fleshy, from the forepart of the body of the first vertebra of the neck, close to the transverse process, and ascending obliquely, is

Inserted near the root of the condyloid process of the occipital bone, under the last muscle: acts as the last.

4. RECTUS CAPITIS LATERALIS—*Arises*, fleshy, from the upper part of the transverse process of the atlas.

Inserted, tendinous and fleshy, into the transverse ridge of the os occipitis.

Situation. It is immediately behind the internal jugular vein, where it comes out from the cranium.

Use. Draws the head somewhat to one side.

CHAPTER VI.

DISSECTION OF THE THROAT.

On looking into the mouth, we observe the tongue, and a soft curtain hanging from the palate bones, named the VELUM PENDULUM PALATI, or soft Palate. The apex of the velum forms a small projecting body, termed the UVULA, or pap of the throat. From each side of the uvula, two muscular half-arches or columns are sent down, the anterior to the root of the tongue, the posterior to the side of the pharynx. Between these half-arches on each side are situated the glands termed Amygdalæ, or Tonsils. The common opening behind the anterior arch is named the Fauces, or top of the

Throat, from which there are six passages, two upward, being one to each nostril, called the Posterior Nares; two at the sides, called Eustachian Tubes, passing on each side to the ear;[1] two downward, of which the anterior is the passage through the glottis and larynx into the trachea; the posterior, which is the largest, is the pharynx, or top of the œsophagus, and leads to the stomach.

Muscles situated about the Entry of the Fauces.

These consist of four pairs, and a single muscle in the middle.

Fig. 23.

THE PHARYNX LAID OPEN AND VIEWED FROM BEHIND.

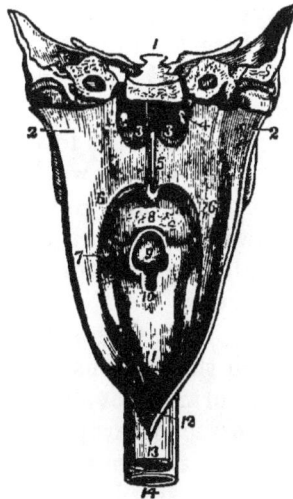

1. A Section carried transversely through the Base of the Skull.
2, 2. The Walls of the Pharynx drawn to each side.
3, 3. The Posterior Nares, separated by the Vomer.
4. The Extremity of the Eustachian Tube of one side.
5. The Soft Palate.
6. The Posterior Pillar of the Soft Palate.
7. Its Anterior Pillar; the Tonsil is seen in the niche between the two Pillars.
8. The Root of the Tongue, partly concealed by the Uvula.
9. The Epiglottis, overhanging (10) the Superior opening of the Larynx.
11. The Posterior Part of the Larynx.
12. The Opening into the Œsophagus.
13. The External Surface of the Œsophagus.
14. The Trachea.

1. CONSTRICTOR ISTHMI FAUCIUM—*Arises*, by a slender beginning, from the side of the tongue, near its root;

---

[1] A probe may be introduced through the anterior nostrils into the Eustachian tube; the tube opens into the pharynx in a direction opposite to the space between the roots of the middle and inferior turbinated bones.

thence running upward within the anterior arch, before
the amygdala, it is

Inserted into the middle of the velum pendulum palati,
as far as the root of the uvula. It is here connected
with its fellow, and with the beginning of the palato-
pharyngeus.

Situation. It forms the anterior half-arch.

Use. To draw the velum toward the root of the
tongue, which at the same time it raises, and, with its
fellow, to contract the opening into the fauces.

2. The PALATO-PHARYNGEUS—*Arises*, by a broad
beginning, from the root of the uvula in the middle of
the velum pendulum palati, and from the tendinous expansion of the circumflexus palati. The fibres pass
along the posterior arch behind the amygdalæ, and run
backward to the superior and lateral part of the pharynx,
where they are scattered, and mixed with those of the
stylo-pharyngeus.

Inserted into the edge of the upper and back part of
the thyroid cartilage, and into the back part of the pharynx.

Situation. It forms the posterior half-arch or column.

Use. To draw the uvula and velum downward and
backward, and pull the thyroid cartilage and pharynx
upward; to shut the passage into the nostrils, and, in
swallowing, to thrust the food from the fauces into the
pharynx.

3. The CIRCUMFLEXUS, or Tensor Palati—*Arises*
from the spinous process of the sphenoid bone, behind
the foramen ovale, and from the Eustachian tube near
its osseous part; runs down along the pterygoideus internus, and forms a round tendon, which passes over the
hook of the internal plate of the pterygoid process of
the sphenoid bone, and soon spreads into a broad tendinous expansion.

Inserted into the velum pendulum palati, and semilunar edge of the os palati.

Situation. Its insertion extends as far as the suture
which joins the two ossa palati. Some of its posterior
fibres generally join with the constrictor pharyngis
superior and palato-pharyngeus.

Use. To stretch the velum, to draw it downward, and to one side.

4. The LEVATOR PALATI — *Arises,* tendinous and fleshy, from the extremity of the petrous portion of the temporal bone, and from the Eustachian tube.

Inserted into the whole length of the velum pendulum palati, as far as the root of the uvula, uniting with its fellow.

Use. To draw the velum upward and backward, so as to shut the passage from the fauces into the mouth and nose.

AZYGOS UVULÆ.—There are two which *arise,* fleshy, from the extremity of the suture which unites the ossa palati; runs down the whole length of the velum, like a small earth-worm, adhering to the tendons of the circumflexi palati.

Inserted into the tip of the uvula.

Use. To raise the uvula upward and forward, and shorten it.

The Tongue.

The tongue is connected by its root to the os hyoides, is principally formed by muscular structure. On its surface note its roughness, formed by papillæ, eight or nine at its posterior part, arranged in a V-shaped figure (PAPILLÆ MAXIMÆ, or *Circumvallatæ*). Smaller and more numerous are others scattered over the tongue (the PAPILLÆ MEDIÆ, or *Fungiformes*). A third class, very numerous (the PAPILLÆ MINIMÆ, or Filiformes). Three folds of mucous membrane pass from the back part of the tongue to the sides and centre of the Epiglottis (GLOSSO-EPIGLOTTIDEAN FOLDS), the middle one sometimes called the FRÆNUM of the EPIGLOTTIS. The Foramen Cæcum of Morgagni is in front of this.

Muscles.

Several of the muscles of the tongue have been described in the dissection of the neck. Those which enter into its structure proper are the

5

LINGUALIS, which *arises* from its root, and runs as far as the tip.

Use. To shorten the tongue.

SUPERFICIALIS LINGUÆ, TRANSVERSALES LINGUÆ, and VERTICALES LINGUÆ.—Their names indicate their direction.

Muscles situated on the Posterior Part of the Pharynx.

Of these there are three pair, and are better dissected after the muscles on the back of the neck are disposed of, in order to remove the head from its articulation with the atlas, taking with it the Pharynx, Larynx, and Œsophagus.

1. The CONSTRICTOR PHARYNGIS INFERIOR. —This muscle *arises* from the outside of the ala of the thyroid cartilage, near the attachment of the thyreo hyoideus muscle, and from the side of the cricoid cartilage, near the crico-thyroideus.

Inserted into the white line on the back part of the pharynx, where it is united to its fellow.

Situation. This muscle covers the under part of the middle constrictor; the superior fibres run obliquely upward, while the inferior fibres have a transverse direction.

Use. To compress that part of the pharynx which it covers, and to raise it with the larynx a little upward.

2. The CONSTRICTOR PHARYNGIS MEDIUS—*Arises* from the superior edge of the cornu of the os hyoides, extending as far forward as the appendix; and from the ligament which connects it to the thyroid cartilage. The superior fibres ascend obliquely, the others run more transversely.

Inserted into the cuneiform process of the os occipitis, before the foramen magnum, and into a white line in the middle of the posterior surface of the pharynx, where it is joined to its fellow.

Situation. The lower part of this muscle is covered by the muscle last described, while the upper part covers the inferior fibres of the constrictor superior.

Use. To compress that part of the pharynx which it invests, and to draw it and the os hyoides upward.

Fig. 24.

A SIDE VIEW OF THE MUSCLES OF THE PHARYNX.

1. The Trachea.
2. The Cricoid Cartilage.
3. The Crico-thyroid Ligament.
4. The Thyroid Cartilage.
5. The Thyro-hyoid Ligament.
6. The Hyoid Bone.
7. The Stylo-hyoid Ligament.
8. The Œsophagus.
9. The Inferior Constrictor Muscle.
10. The Middle Constrictor Muscle.
11. The Superior Constrictor Muscle.
12. The Stylo-pharyngeus Muscle passing down between the Superior and Middle Constrictors.
13. The Upper Concave Border of the Superior Constrictor. At this point the muscular fibres of the Pharynx are deficient.
14. The Pterygo-Maxillary Ligament.
15. The Buccinator Muscle.
16. The Orbicularis Oris Muscle.
17. The Mylo-hyoid Muscle.

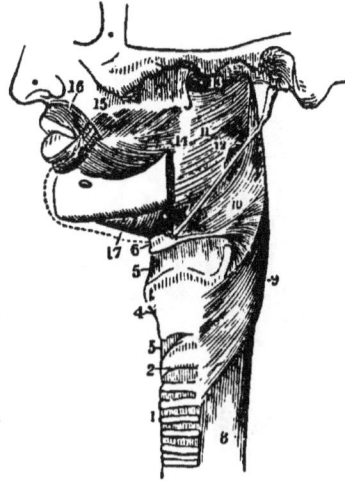

3. CONSTRICTOR PHARYNGIS SUPERIOR — *Arises,* above, from the cuneiform process of the os occipitis, before the foramen magnum; lower down from the pterygoid process of the sphenoid bone; from the upper and under jaw, near the alveolar process of the last dentes molares; and from the back part of the buccinator muscle. Some fibres also come from the root of the tongue, and from the palate.

Inserted into a white line in the middle of the posterior surface of the pharynx.

Situation. The larger part of this muscle is covered by the constrictor medius.

Use. To compress the upper part of the pharynx and draw it forward and upward.

Muscles of the Larynx.

The Larynx is composed of nine cartilages: 1. The THYROID Cartilage, situated immediately below the os hyoides in the middle of the throat. 2. The CRICOID Cartilage, situated immediately below the thyroid cartilage, betwixt it and the superior rings of the trachea. 3. The EPIGLOTTIS, a broad triangular cartilage, very

Fig. 25.

CARTILAGES OF THE LARYNX SEPARATED AND SEEN IN FRONT.

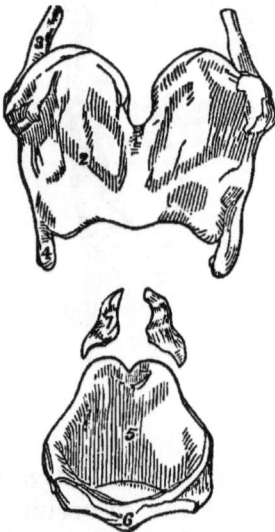

1 to 4. Thyroid Cartilage.
    1. Vertical Ridge, commonly called Adam's Apple, formed by the union of the two Plates or Halves.
    2. Right Half.
    3. Superior, and
    4. Inferior Horn of the Right Side.
5, 6. Cricoid Cartilage.
    7. Right Arytenoid Cartilage.

elastic, situated behind the root of the tongue, and covering the entrance into the upper part of the larynx. 4 and 5. The ARYTENOID Cartilages, two small bodies, like peas, situated behind the thyroid cartilage, on the upper edge of the back part of the cricoid cartilage, and between the two alæ or wings of the thyroid cartilage. These two small cartilages form betwixt themselves and the thyroid a longitudinal fissure, extending from before backward, which is called the Glottis, or Rima Glottidis, and leads to the trachea. 6 and 7. CORNICULA Laryngis. These surmount the arytenoids.

8 and 9. CUNEIFORM Cartilages (Cartilages of Wrisberg) in the Aryteno-Epiglottidean folds.

The muscles situated about the glottis consist of four pair of small muscles, and a single one.

1. The CRICO-ARYTÆNOIDEUS POSTICUS—*Arises*, fleshy, from the posterior part of the cricoid cartilage, and is

Inserted, narrow, into the back part of the arytenoid cartilage of the same side.

Use. To open the rima glottidis a little, and, by pulling back the arytenoid cartilage, to render the ligament of the glottis tense.

2. The CRICO-ARYTÆNOIDEUS LATERALIS—*Arises*, fleshy, from the side of the cricoid cartilage, where it is covered by the ala of the thyroid cartilage.

Inserted into the outer side of the arytenoid cartilage.

Situation. It lies more forward than the last described muscle.

Use. To open the rima glottidis, by pulling the ligaments from each other.

3. The THYREO-ARYTÆNOIDEUS—*Arises* from the middle and inferior part of the posterior surface of the thyroid cartilage; runs backward, and a little upward, and is

Inserted into the forepart of the arytenoid cartilage.

Situation. It is situated more forward than the muscle last described.

Use. To pull the arytenoid cartilage forward, and thus shorten the ligament of the larynx or glottis.

ARYTÆNOIDEUS OBLIQUUS—*Arises* from the base of one arytenoid cartilage; and, crossing its fellow, is

Inserted into the tip of the other arytenoid cartilage.

Use. When both act, they pull the arytenoid cartilages toward each other.

The single muscle is the

ARYTÆNOIDEUS TRANSVERSUS, which *arises* from the whole length of one arytenoid cartilage, and passes across, to be

Inserted into the whole length of the other arytenoid cartilage.

Fig. 26.

LARYNGEAL MUSCLES.

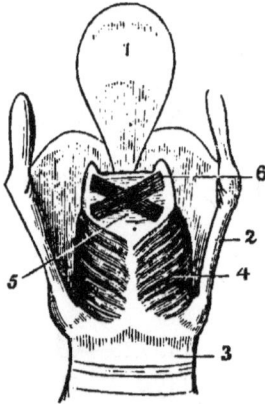

1. Epiglottis.
2. Thyroid Cartilage.
3. Cricoid Cartilage.
4. Crico-arytenoideus Posticus.
5. Arytenoideus Transversus.
6. Arytenoideus Obliquus.

Situation. It is situated anterior to the arytænoidei obliqui.

Use. To shut the rima glottidis, by bringing the arytenoid cartilages together.

Fig. 27.

MUSCLES AND CARTILAGES OF THE LARYNX.

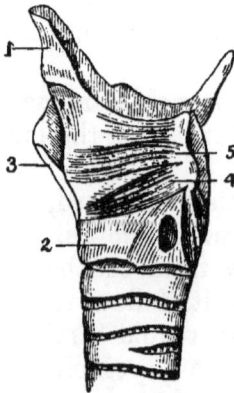

1. Epiglottis.
2. Cricoid Cartilage.
3. Thyroid Cartilage.
4. Crico-arytenoideus Lateralis.
5. Thyro-arytenoideus.

On each side of the larynx there are also a few muscular fibres, which are named as follows:

1. THYREO EPIGLOTTIDEUS—*Arising*, by a few pale separated fibres, from the thyroid cartilage, and

Inserted into the epiglottis laterally.

Use. To draw the epiglottis obliquely downward, or, when both muscles act, directly downward; and at the same time to expand it.

2. The ARYTÆNO-EPIGLOTTIDEUS—*Arises*, by a few slender fibres, from the lateral and upper part of the arytenoid cartilage, and is

Inserted into the epiglottis, along with the former muscle.

Use. When both muscles act, to pull the epiglottis close upon the glottis.

The Thyreo-Hyoideus and the Crico-Thyroideus were described with the muscles of the neck. The inside of the larynx is lined with a mucous membrane, and two folds will be seen running from the arytenoid cartilages to the angle of the thyroid; on either side these are the VOCAL CORDS. Between the cords of each side a cavity (the VENTRICLE), which communicates with another cavity at the upper and front part (the SAC of HILTON, or SACCULUS LARYNGIS). The cavity above the upper cords (the glottis); space between the lower cords, the RIMA GLOTTIDIS.

CHAPTER VII.

DISSECTION OF THE ORBIT OF THE EYE.

THE globe or ball of the eye is situated about the middle of the orbit. It is connected to the bone by its muscles, and by the optic nerve; and all these parts are imbedded posteriorly in a soft, fatty substance, which fills up the bottom of the orbit. The tunica, or membrana conjunctiva, is seen lining the inner surface of the eyelids, and reflected from them over the anterior part of the globe of the eye, so that the back part of the eyeball, and all the muscles and nerves, are situated behind it. This membrane must therefore be dissected away, the upper part of the orbit, which is formed by the os frontis, removed with a saw, and the fat surrounding the muscles, vessels, and nerves, cautiously dissected away with the scissors.

Muscles situated within the Orbit.

Seven muscles are contained within the orbit, of which one belongs to the upper eyelid, and six to the globe of the eye.

1. The LEVATOR PALPEBRÆ SUPERIORIS—*Arises*, by a small tendon, from the upper part of the foramen opticum of the sphenoid bone; the tendon forms a broad flat belly.

Inserted, by a broad thin tendon, into the upper eyelid, adhering to the tarsal cartilage, which gives form to the eyelid.

Use. To open the eye, by drawing the superior eyelid upward.

There are four straight muscles, or recti, belonging to the globe of the eye. These four muscles resemble each other, all arising by narrow tendons from the margin of the foramen opticum, where they surround the optic nerve; all forming strong fleshy bellies, and inserted, by broad, thin tendons, at the forepart of the globe of the eye, into the tunica sclerotica, or outer tunic of the eye, and under the tunica conjunctiva.

2. The LEVATOR OCULI, or Rectus Superior—*Arises*, by a narrow tendon, from the upper part of the foramen opticum of the sphenoid bone; it forms a fleshy belly, and is

Inserted into the superior and anterior part of the tunica sclerotica, by a broad thin tendon.

Situation. It lies below the levator palpebræ superioris.

Use. To raise the globe of the eye.

3. The DEPRESSOR OCULI, or Rectus Inferior—*Arises* from the inferiôr margin of the foramen opticum, and is

Inserted into the inferior and anterior part of the tunica sclerotica.

Use. To move the globe of the eye downward.

4. The ADDUCTOR OCULI, or Rectus Internus—*Arises* from the inner part of the foramen opticum, and is

Inserted into the inner and anterior part of the tunica sclerotica.

It is the shortest of the four recti muscles.

Use. To draw the eye toward the nose.

5. The ABDUCTOR OCULI, or Rectus Externus—*Arises* from the outer part of the foramen opticum.

Inserted into the outer part of the tunica sclerotica.

It is the longest of the recti.

Use. To move the globe outward.

The two next are oblique muscles.

Fig. 28.

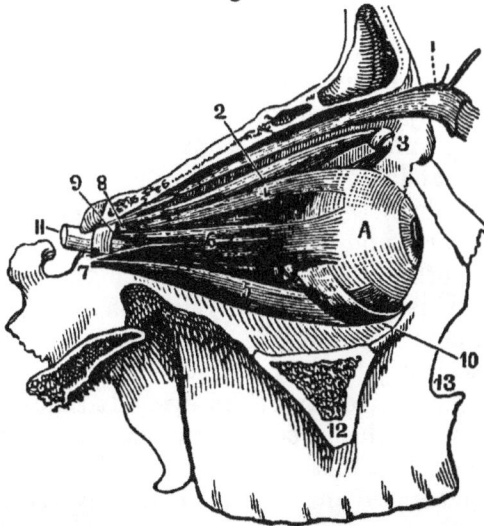

A SIDE VIEW OF THE MUSCLES OF THE EYEBALL.

A. Ball of the Eye.
1. Levator Palpebræ Superioris.
2. Obliquus Superior, or Trochlearis Muscle.
3. Trochlea of the last named Muscle.
4. Rectus Superior Muscle.
5. Rectus Inferior Muscle.
6. Rectus Externus Muscle.
7. Ligament of Zinn.
8. Origin of the Superior Oblique Muscle.
9. Origin of the Rectus Externus.
10. Obliquus Inferior Muscle.
11. Optic Nerve.
12 Malar Bone, divided.
13. Upper Maxillary Bone.

6. The OBLIQUUS SUPERIOR, or Trochlearis—*Arises,* by a small tendon, from the margin of the foramen opticum, between the levator and abductor oculi. Its long slender belly runs along the inner side of the orbit to

5*

the internal angular process of the os frontis, where a cartilaginous pulley is fixed. The muscle then forms a tendon, which passes through the pulley, runs obliquely downward and outward, inclosed in a membranous sheath; and, becoming broader and thinner, is

Inserted into the tunica sclerotica, about half way between the insertion of the levator oculi and the entrance of the optic nerve.

Use. To roll the globe of the eye, and turn the pupil downward and outward.

7. The OBLIQUUS INFERIOR—*Arises,* narrow, and principally tendinous, from the outer edge of the orbitar process of the superior maxillary bone, near its junction with the os unguis. It runs obliquely outward and backward, under the depressor oculi, and is

Inserted, by a broad thin tendon, into the sclerotica, between the entrance of the optic nerve and the insertion of the abductor oculi.

Use. To draw the globe of the eye forward, inward, and downward, and to turn the pupil upward.

In the orbit we also meet with the LACHRYMAL GLAND. This gland is of a yellowish color; it is situated in a depression of the os frontis near the temple. It adheres closely to the fat which surrounds the muscles and posterior convexity of the eye. It sends off several small ducts, which pierce the tunica conjunctiva lining the upper eyelid; these ducts cannot be seen, unless the part be macerated in water, when they are filled with the liquid.

The PALPEBRÆ, or Eyelids, are two cartilaginous plates, semilunar in form, the upper the largest, connected to the internal and external angle of the orbit by fibrous tissue called the external and internal palpebral ligament (the internal one sometimes called the TENDO OCULI), above and below to the periosteal margin of the orbit by a fibrous membrane.

The PUNCTA LACHRYMALIA are two small holes near the internal angle of the palpebræ, situated one in each eyelid. They lead into the LACHRYMAL DUCTS, the ducts into the lachrymal sac.

The LACHRYMAL SAC is an oblong membranous bag, situated at the inner angle of the eye, in a depression formed by the os unguis and nasal process of the superior maxillary bone. It receives the tears by the puncta lachrymalia, and from the sac they are conveyed into the nose by a DUCT, named the LACHRYMAL or NASAL. The lower extremity of this duct opens into the nose on one side of the antrum maxillare, and under the os spongiosum inferius. A probe, with its extremity bent, may be introduced from the nose through this duct into the lachrymal sac.

The Caruncula Lachrymalis is a small reddish granulated body, situated at the internal angle of the palpebræ.

TENSOR TARSI of Horner—*Arises* from the posterior superior surface of the os unguis, passes forward and outward, lying on the posterior face of the lachrymal ducts, upon which it is inserted nearly as far as the puncta lachrymalia.

Use. Draws the ducts toward the eye.

Of the Vessels and Nerves met with in the Orbit of the Eye.

Arteries.

The OPHTHALMIC or OCULAR ARTERY is a branch of the internal carotid. It enters the orbit from the basis of the cranium by the foramen opticum. It gives branches to the lachrymal gland, fat, muscle, and globe of the eye. One twig, named the A. CENTRALIS RETINÆ, enters the substance of the optic nerve, and is continued on to the retina; twigs also pass to the eyelids and to the inner angle of the eye. The ARTERIA FRONTALIS is a branch of this artery; it is seen running toward the supra-orbitary notch or foramen, and is distributed to the forehead.

The INFRA-ORBITARY ARTERY is found in the lower part of the orbit; it is the continued trunk of the internal maxillary, entering the orbit by the spheno-maxillary slit. It is seen passing along the canal in the upper part of the great tuberosity of the os maxillare superius, and emerges on the face by the infra-orbitary hole.

Veins.

These correspond with the arteries; they discharge their blood partly into the branches of the external jugular vein near the forehead and temples, and partly into the internal jugular.

Nerves.

1. The OPTIC NERVE is seen coming through the foramen opticum, and entering the back part of the globe of the eye, to form the retina.

2. The Nerve of the Third Pair, MOTOR OCULI, having entered the orbit through the superior orbitary fissure, or foramen lacerum, is divided into four branches.

(1) The first runs upward, and subdivides into two nerves, of which one supplies the superior rectus, and the other the levator palpebræ superioris.

(2) The second branch goes to the superior rectus, and is short.

(3) The third branch supplies the obliquus inferior, and also gives off a twig, which assists in forming the lenticular ganglion.

(4) The fourth branch supplies the internal rectus.

3. The Nerve of the Fourth Pair, N. PATHETICUS or Trochlearis, enters the orbit by the superior orbitary fissure, and runs to the obliquus superior.

4. The first branch of the Nerve of the Fifth Pair, named OPHTHALMIC or Orbitary, enters the orbit by the superior orbitary fissure, and divides into three branches.

(1) The FRONTAL, Supra-orbitary or Superciliary Nerve, accompanies the frontal artery along the upper part of the orbit, close to the bone; and having passed through the supra-orbitary notch, is distributed to the forehead.

(2) The NASAL Nerve, or inner branch, runs toward the nose, and is distributed to the inner side of the orbit and to the nose.

(3) The Temporal or Lachrymal Branch supplies the lachrymal gland and the parts at the outer side of the orbit.

The LENTICULAR GANGLION is a small ganglion, situated within the orbit, formed by short branches of the ophthalmic nerve, and by a twig of the third pair. It sends off delicate nerves, which run along the sides of the optic nerve, and pierce the coats of the eye.

5. The second branch of the Fifth Pair, called the SUPERIOR MAXILLARY NERVE, sends off a branch through the bony canal in the bottom of the orbit. This is the INFRA-ORBITARY NERVE. It accompanies an artery of the same name, and emerges on the face by the INFRA-ORBITARY FORAMEN.

6. The trunk of the Sixth Pair of nerves passes through the sphenoidal fissure to the external rectus muscle.

These delicate nerves are surrounded with fat, and demand great care in their dissection.

CHAPTER VIII.

DISSECTION OF THE THORAX.

Of the Muscles which lie upon the Outside of the Thorax. The Axilla.

To expose these muscles, carry an incision from the top of the sternum along the median line to the ensiform cartilage, from thence obliquely upward through the axilla, and a few inches down the inner side of the arm. From the commencement of the first another along the clavicle to the acromion process.

Notice the axilla, a conical cavity, having a muscular fold in front and behind. The former made by the PEC-

TORALIS MAJOR muscle, the latter by the LATISSIMUS
DORSI muscle.

In removing the integuments from the forepart of the
thorax, the pectoralis major and interior edge of the deltoid muscle should be dissected in the course of their
fibres; and to do this, it will be necessary to remember
that the fibres run obliquely from the sternum and clavicle to the upper part of the os humeri.

Of the FEMALE MAMMÆ. — Two glandular bodies
placed upon the great pectoral muscles, having only the
fascia interposed, upon which they readily move. In the
centre is placed the NIPPLE, about which are many sebaceous follicles. Ten or fifteen ducts (TUBULI LACTI-
FERI) commence at its extremity, and passing inward
divide and subdivide, which ultimately end in enlargements, from which pass off many others, to end finally
in the ultimate lobules or vesicles of the organ. The
lobules are supported by prolongations from the cellular
capsule of the gland, which pass through its substance.
Much adipose structure is usually intermixed with the
lobules.

Its bloodvessels from the axillary, intercostal, and internal mammary trunks.

Three pair of muscles are described in the dissection
of the thorax.

1. The PECTORALIS MAJOR—*Arises*, tendinous, from
the anterior surface of the sternum, its whole length;
fleshy, from the cartilages of the fifth, sixth, and sometimes the seventh ribs, and from two anterior thirds of
the clavicle. The fleshy fibres run obliquely across the
breast, and, converging, form a strong, flat tendon,
which is

Inserted into the ridge of the os humeri on the outside of the bicipital groove.

Situation. The belly of this muscle is superficial. It
is separated from the deltoid muscle by cellular membrane and fat, by the CEPHALIC VEIN, and a small
artery, named HUMERAL THORACIC. Its tendinous fibres,
arising from the sternum, are interlaced with those of
the opposite, so as to form a kind of fascia covering the

bone; and the origins from the ribs are intermixed with the obliquus externus abdominis. The clavicular and thoracic portions of the muscle are separated by a line of cellular membrane which leads to the axillary vessels.

Fig. 29.

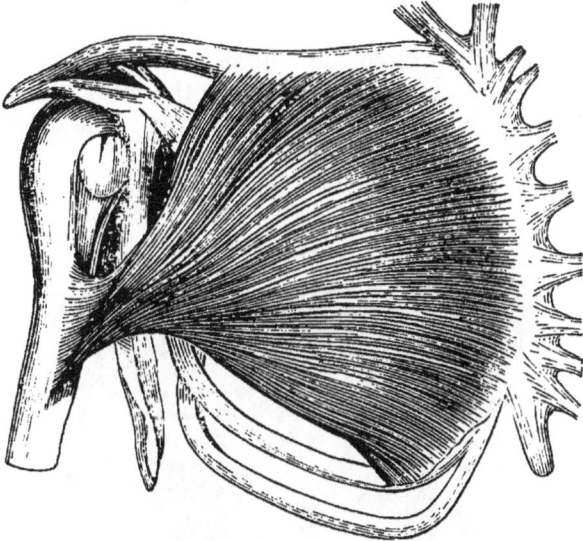

The tendon is covered by the anterior edge of the deltoid; it forms the anterior fold of the armpit, and appears twisted, for the fibres which proceed from the thoracic portion of the muscle seem to pass behind those proceeding from the clavicle, and to be inserted into the os humeri somewhat higher up.

Use. To move the arm forward, and obliquely upward, toward the sternum.

The pectoralis major should be lifted up from its origin. This will expose the next two muscles.

2. The PECTORALIS MINOR—*Arises,* by three tendinous and fleshy digitations from the upper edges of the third, fourth, and fifth ribs, near their cartilages; it forms a fleshy triangular belly, which becomes thicker and narrower as it ascends, and is

Inserted, by a short flat tendon, into the anterior part of the coracoid process of the scapula.

Situation. The belly of this muscle is covered by the pectoralis major: the tendon passes under the anterior edge of the deltoid, and is connected at its insertion with the origins of the coraco-brachialis, and of the short head of the biceps flexor cubiti, and also with the coraco-acromial ligament.

Use. To draw the scapula forward and downward, and, when that bone is fixed, to elevate the ribs. Above its insertion notice some arteries presently to be described.

Fig. 30.

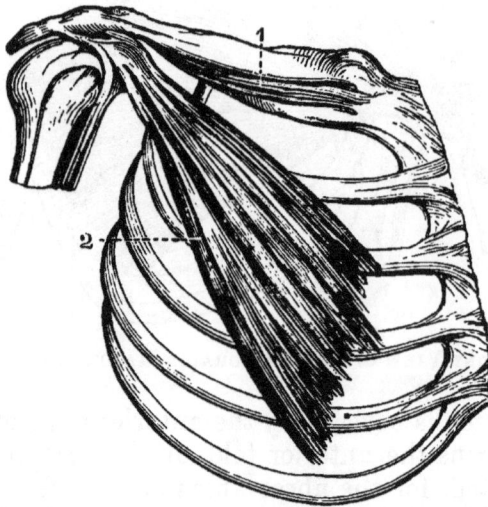

A FRONT VIEW OF THE SUBCLAVIUS AND PECTORALIS MINOR MUSCLES.

1. Subclavius.   2. Pectoralis Minor.

3. The SUBCLAVIUS—*Arises*, by a flat tendon, from the cartilage of the first rib, and forms a broad fleshy belly, which is

Inserted into the inferior surface of the clavicle, beginning about one inch from the sternum, and continuing as the ligamentous connection of the clavicle to the coracoid process.

Situation. This muscle is situated between the clavicle and sternum, concealed by the pectoralis major, and anterior part of the deltoides.

Use. To draw the clavicle downward and forward, and perhaps to elevate the first rib.

Having lifted up the pectoralis minor from its origin, the situation of the subclavian vessels which pass under the clavicle, and over the first rib, may be seen.

Dissection of the Axilla.

The axilla is formed by two muscular folds, which bound a middle cavity. The anterior fold is formed by the pectoralis major passing from the thorax to the arm, the posterior by the latissimus dorsi coming from the back.

In the intermediate cavity there is a quantity of cellular membrane and absorbent glands and fat, covering and connecting the great vessels and nerves; and this is not exactly similar to the fat in other parts of the body; it is more granulated, watery, and of a reddish color; it surrounds the great vessels and nerves, rendering the dissection both tedious and difficult. The lymphatic glands here are continuous under the clavicle with those of the neck.

The AXILLARY VEIN will be found lying anterior to the artery, that is, nearer the integuments. The axillary vein receives branches corresponding to the ramifications of the artery. Passing under the clavicle, it becomes the subclavian vein, and runs over the first rib, and before the anterior scalenus muscle into the thorax.

Deeper seated, and immediately behind the axillary vein, lies the AXILLARY ARTERY. It is seen coming from under the clavicle; from under the arch formed by the pectoralis minor, it comes over the middle of the first rib, and between the anterior and middle scaleni muscles. In the axilla it is surrounded by the meshes of the axillary nerves, and runs under the tendon of pectoralis minor along the inferior edge of the coracobrachialis muscle; when it has passed to the tendon of the latissimus dorsi muscle, it assumes the name of the Brachial Artery.

The branches of the axillary artery are

1. A. MAMMARIA EXTERNA, called also A. Thoracicæ Externæ.—The external mammary artery consists of four or five branches which run downward and forward obliquely over the chest. These branches sometimes come off separately from the axillary artery, at other times by one or two common trunks, which subdivide. They supply the pectoral muscles and mamma. Some of their branches pass to the muscles of the shoulder, to the side of the chest, and to the muscles on the inside of the scapula. They are as follows:

a. THORACICO ACROMIALIS—going toward the fissure between the deltoid and pectoralis major.

b. THORACICO SUPERIOR—to the pectoralis major.

c. THORACICO LONGA—along the border of the pectoralis minor and side of the chest.

d. THORACICO AXILLARIS—to the glands and cellular tissue of the armpit.

2. A. SUBSCAPULARIS *arises* from the under and back part of the axillary artery, attaches itself to the inferior costa of the scapula, where it splits into two great branches: 1. The Dorsalis Scapula, a large branch, which passes to the outer surface of the scapula below the spine, and has its principal ramifications close upon the bone. 2. The other branch (which is larger) passes to the inner surface of the scapula, supplies the subscapularis, and sends branches downward to the muscles of the back and loins. (See Fig. 30.)

3. ARTERIA CIRCUMFLEXA HUMERI POSTERIOR *arises* from the lower and forepart of the axillary artery, and runs backward close to the bone, surrounds its neck, and is lost on the inner surface of the deltoid; it gives also twigs to the joint and neighboring muscles. It is accompanied by the CIRCUMFLEX NERVE.

4. A. CIRCUMFLEXA ANTERIOR is a much smaller artery, often a branch of the circumflexa posterior; it encircles the neck of the bone on its forepart, and is

lost on the inner surface of the deltoides, where it inosculates with the posterior circumflex artery.

The GREAT BRACHIAL NERVES accompany the subclavian artery over the first rib. In the axilla they are united by numerous cross branches, forming the Axillary or Brachial Plexus, which is continued from the clavicle as low as the edge of the tendon of the latissimus dorsi, and which surrounds the axillary artery with its meshes.

From the axillary plexus seven nerves pass off.

1. NERVUS SUPRA-SCAPULARIS.—This nerve comes off from the upper edge of the plexus; it crosses the axilla at the highest part, runs toward the superior costa of the scapula, accompanies the external scapular artery through the semilunar notch, and supplies the muscles on the posterior surface of the scapula.

2. N. CIRCUMFLEXUS lies deep; it passes from the back part of the plexus, goes backward round the neck of the bone, accompanying the posterior circumflex artery, and is distributed to the deltoid and the muscles on the outside of the arm. Small nerves also pass from the axillary plexus to the subscapular muscle (subscapular), the teres major, latissimus dorsi, and pectoral muscles.

3. The External Cutaneous Nerve, or Nervus Musculo-cutaneus.

4. The Median Nerve.

5. The Ulnar Nerve.

6. The Musculo-Spiral Nerve.

7. The Internal Cutaneous Nerve.

8. The Lesser Internal Cutaneous, or Nerve of Wrisberg, formed by a filament from the second intercosto humeral, and another from the axillary plexus.

At this exposure of parts a nerve may be observed descending close along the thorax upon the serratus magnus muscle. This is the LONG THORACIC or EXTERNAL RESPIRATORY of Bell, and comes from the fourth and fifth cervical nerves. Two others come through the intercostal spaces, the INTERCOSTO HUMERAL. They come from the intercostal nerves, and mingle with the axillary nerves.

Six of these nerves will be described in the dissection of the arm and forearm.

- - - - -

CHAPTER IX.

DISSECTION OF THE SUPERIOR EXTREMITY.

Of the Shoulder and Arm.

AN incision carried down the middle of the arm and integuments reflected, will expose the parts properly for dissection. The arm is invested with a superficial fascia, mingled with fat, in and beneath which are situated important veins. Beneath the superficial fascia is the DEEP FASCIA, consisting of fibres running circularly and longitudinally. This fascia is very thin over the deltoid. Strong at the internal condyle, and on the back of the forearm, sending processes in between the muscles and furnishing to them a surface of origin.

In removing the integuments, we meet with several cutaneous veins and nerves.

The cutaneous veins[1] of the upper extremity are the following:

1. The BASILIC VEIN is seen arising from a small vein on the outside of the little finger, named Salvatella. It then runs along the inside of the forearm near the ulna, receiving the internal and external ulnar veins from the anterior and posterior surface of the fascia. It passes over the fold of the arm near the inner condyle of the humerus. It ascends along the arm, becoming more deeply seated, and included in the sheath which invests the brachial artery. As it approaches the neck

- - - - -

[1] The veins are described from their origin in the forearm for the sake of perspicuity.

of the humerus, it sinks deep betwixt the folds of the armpit, and terminates in the axillary vein, which may be considered as a continuation of the basilic vein. It communicates with the deeper-seated veins, and receives numerous branches from the muscles.

Fig. 31.

SUPERFICIAL VEINS OF THE SUPERIOR EXTREMITY.

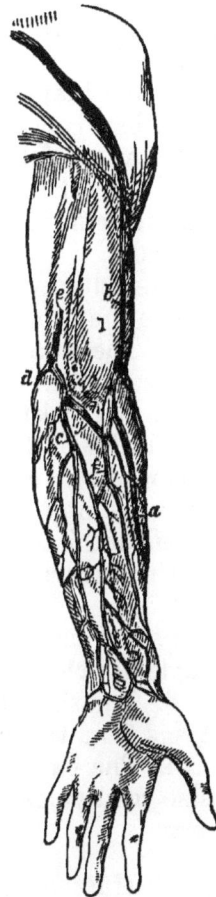

a. Commencement of the Cephalic Vein.
b. Main Trunk of Cephalic Vein.
c. Anterior Branch of Basilic Vein.
d. Posterior Branch of Basilic Vein.
e. Basilic Vein.
f. Median Vein.
g. Median Basilic Vein.
h. Median Cephalic Vein.
i. Biceps Muscle.

2. The CEPHALIC VEIN begins on the back of the hand, between the thumb and metacarpal bone of the forefinger, by a small vein, named CEPHALICA POLLICIS. It runs along the radius between the muscles and in-

teguments, receiving the internal and external radial
veins.   It passes over the bend of the arm near the external condyle, and ascends along the outside of the arm
near the outer edge of the biceps flexor cubiti.   It then
runs betwixt the edge of the deltoid and pectoral muscles, dips down under the clavicle, and enters the subclavian vein.   In all this course the cephalic vein receives
branches.

3. The MEDIAN VEIN.—Several veins are seen running along the middle of the anterior part of the forearm.   The trunk formed by these veins is called the
Mediana Major.   It ascends on the flat part of the forearm, betwixt the basilic and cephalic veins, and bifurcates at the fold of the arm into two branches: 1. The
Mediana Basilica, passing off obliquely to join the
basilic vein; 2. The Mediana Cephalica, which joins the
cephalic.

The cutaneous nerves of the arm are seen ramifying
above the muscles; they consist of

1. The Internal Cutaneous Nerve, a branch of the
axillary plexus.   It is seen accompanying the basilic
vein, and twisting its fibres over it.   It descends along
the inside of the arm, crosses over the forepart of the
elbow-joint, and, in the dissection of the forearm, will be
seen dividing itself into twigs, which ramify between the
fascia and integuments, and are distributed to the inside
of the forearm and wrist.

2. The upper part of the arm receives cutaneous
nerves from the branches of the dorsal nerves, which
come out of the thorax between the ribs.

3. The shoulder and back part of the scapula receive
twigs from the cervical nerves.

4. The external cutaneous, ulnar, and spiral nerves,
also send twigs to the integuments of the arm and forearm.

Muscles situated on the Shoulder and Arm.

These are ten in number.

1. The DELTOIDES—*Arises*, tendinous and fleshy,
from the posterior third of the clavicle, from the whole

edge of the acromion, and from the lower margin of the whole spine of the scapula. From these several origins the fibres run in different directions, and converge to be

Inserted, tendinous, into a triangular rough surface on the outer side of the os humeri, near its middle.

Fig. 32.

A VIEW OF THE DELTOID MUSCLE.

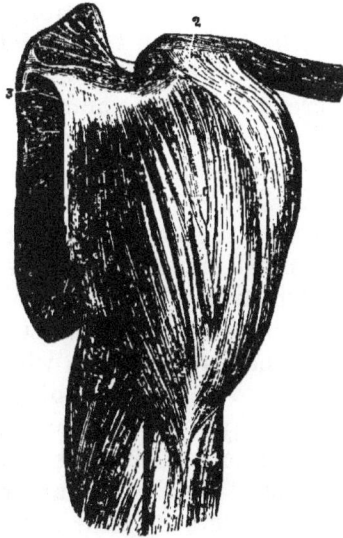

1. Body of the Muscle.
2. Its Insertion into the Clavicle.
3. Its insertion into the Spine of the Scapula.
4. Its Insertion into the Humerus.

Situation. This muscle is entirely superficial, except where the thin fibres of the platysma myoides arise from its anterior surface. It arises from the same extent of bone as the trapezius is inserted into. It is a coarse muscle, consisting of large fasciculi of fibres. It conceals the insertion of the pectoralis major, and the origins of the biceps flexor cubiti and coraco-brachialis, and covers the whole of the forepart and outside of the shoulder-joint. Its external surface is quite fleshy; but, on cutting it across, its internal surface is found tendinous; and where it slides over the great tuberosity of the humerus there is a large bursa.

From the insertion of the deltoid to the outer condyle of the os humeri is extended an Intermusculai Ligament, which separates the muscles on the anterior part of the arm from those on the posterior part, and gives attachment to the fibres of both. It is named the External Intermuscular Ligament. It is only the deep fascia.

Use. To draw the arm directly upward, and to move it a little forward and backward, according to the different directions of its fibres. Reflect it from the scapula and clavicle, that you may expose more completely the muscles on the dorsum of the former bone.

The following two muscles, which fill up the posterior surface of the scapula, are covered by a fascia, which adheres to the spine and edges of that bone. On dissecting off this fascia, the fleshy fibres of the muscles will be found arising from its inner surface.

2. The SUPRA-SPINATUS—*Arises*, fleshy, from all that part of the base of the scapula that is above its spine, from the superior costa as far forward as the semilunar notch, from the spine itself, and from the concave surface betwixt it and the superior costa. The fleshy fibres terminate in a tendon which passes under the acromion, slides over the neck of the scapula (to which it is connected by loose cellular membrane), adheres to the capsular ligament of the shoulder-joint, and is

Inserted into the anterior and superior part of the great tuberosity near the head of the os humeri.

Situation. This muscle fills up the fossa or cavity above the spine of the scapula, and is entirely concealed by the fibres of the trapezius.

Use. To raise the arm.

The supra-scapular artery and nerve pass through this fossa.

3. The INFRA-SPINATUS—*Arises*, principally fleshy, from the lower part of the spine of the scapula as far back as the triangular flat surface; from the base of the bone below the spine to near the inferior angle; from the posterior ridge of the inferior costa; and from all the dorsum of the bone below the spine. The fibres ascend and descend toward a middle tendon, which runs for-

ward over the neck of the bone, and adheres to the capsular ligament.

Inserted, by a strong short tendon, into the middle part of the great tuberosity of the os humeri.

Situation. This muscle is in part concealed by the deltoid, and the trapezius passes over its upper and back part; but a considerable portion of the belly of this

Fig. 33.

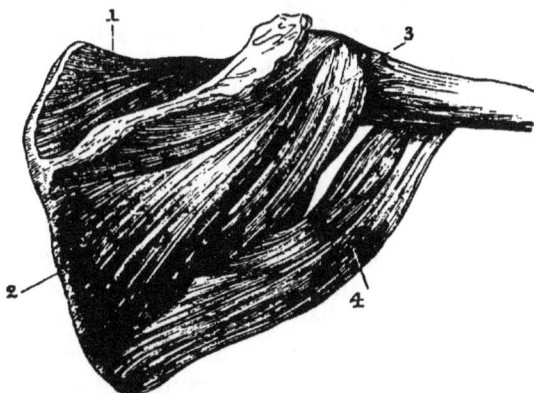

MUSCLES OF THE SCAPULA.

1. Supra-spinatus.
2. Infra-spinatus.
3. Teres Minor.
4. Teres Major.

muscle is seen betwixt these two muscles. It is inserted below the tendon of the supra-spinatus.

Use. To roll the humerus outward, to assist in raising the arm, and in moving it outward when raised.

4. The TERES MINOR—*Arises*, fleshy, from the narrow depression between the two ridges in the inferior costa of the scapula, extending from the neck of the bone to within an inch or two of the inferior angle. It passes forward along the inferior edge of the infra-spinatus, adheres to the capsular ligament of the shoulder-joint, and is

Inserted, tendinous and fleshy, into the lower and back part of the great tuberosity of the os humeri.

Situation. It is inserted below the tendon of the infra-

6

spinatus. Its origin lies between the infra-spinatus and
teres major, and partly concealed by them. Its insertion is concealed by the deltoid. The fascia which
covers the infra-spinatus envelops also the teres minor;
and the two muscles are in some subjects so closely
united as to be with difficulty separated.

Use. To draw the humerus downward and backward,
and to roll it outward.

5. The TERES MAJOR—*Arises* from an oblong, rough,
flattened surface, at the inferior angle of the scapula. It
forms a thick belly, which passes forward and upward
toward the inside of the arm.

Inserted, by a broad thin tendon, into the ridge of the
os humeri, at the inner side of the bicipital groove.

Situation. Its belly passes before the long head of
the triceps extensor cubiti. Its tendon is inserted along
with the tendon of the latissimus dorsi. Observe the
relative situation of these tendons; they both pass under
the coraco-brachialis and short head of the biceps flexor,
to reach the place of their insertion. They appear at
first inseparably united, but on dividing them with some
care we find an intermediate cavity lubricated with
synovia.

Use. To roll the humerus inward, and to draw it
backward and downward.

6. The SUBSCAPULARIS—*Arises,* fleshy, from all the
base of the scapula internally, from the superior and
inferior costæ, and from the whole internal surface of
the bone. It consists of tendinous and fleshy bundles,
which converge, slide over the inner surface of the neck
of the scapula, pass in the hollow under the root of the
coracoid process, and adhere to the inner part of the
capsular ligament of the shoulder-joint.

Inserted, by a strong tendon, into the lesser tuberosity near the head of the os humeri.

Situation. The whole of this muscle is concealed by
the scapula and muscles of the shoulder. It lies betwixt
that bone and the serratus magnus.

Use. To roll the os humeri inward, and to draw it to
the side of the body.

7. The BICEPS FLEXOR CUBITI—*Arises* by two heads. The first and outermost, called the Long Head, arises, by a strong tendon, from a smooth surface in the upper

Fig. 34.

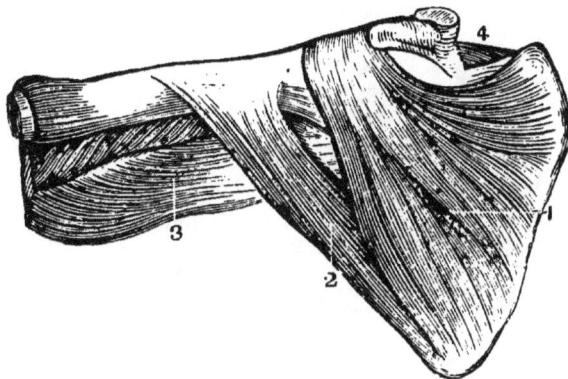

MUSCLES OF THE SCAPULA.

1. Subscapularis.
2. Teres Major.
3. Part of Triceps.
4. Deltoid Muscle.

edge of the glenoid cavity of the scapula. It passes over the head of the os humeri, within the capsular ligament of the shoulder joint, and enters the bicipital groove. It forms a strong fleshy belly. The second and innermost, called the Short Head, arises, tendinous, from the lower part of the coracoid process of the scapula, in common with the coraco-brachialis, and sends off a fleshy belly.

These two fleshy bellies form a thick mass, and, below the middle of the arm, become inseparably united. They send off a strong tendon, which passes into the forepart of the elbow-joint.

Inserted into the posterior rough part of the tubercle of the radius. A bursa mucosa is placed between the tendon and front of the tubercle.

Situation. The tendon of the long head cannot be seen till the capsular ligament of the shoulder is opened. The two origins are concealed by the deltoides and pectoralis major; and, at the bend of the elbow, its tendon sends off from its inside an aponeurosis (BICIPITAL), which assists in forming the fascia of the forearm, and covers the brachial artery and median nerve.

Fig. 35.

MUSCLES OF THE ANTERIOR BRACHIAL REGION, THE ANTERIOR HALF OF THE DELTOID BEING CUT AWAY.

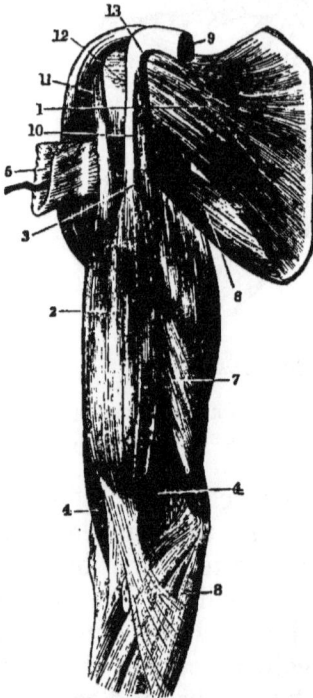

1. Subscapularis Muscle.
2. Biceps.
3, 6. Teres Major.
4, 4. Brachialis Anticus.
5. Tendon of the Pectoralis Major.
7. Internal Head of the Triceps.
8. Tendinous Expansion of the Biceps.
9. Extremity of the Pectoralis Minor.
10. Coraco Brachialis.
11. Long Head of the Biceps.
12. Short Head of the Biceps.
13. Coracoid Process of the Scapula.

Use. To turn the hand supine, to bend the forearm on the arm, and the arm on the shoulder.

8. The CORACO-BRACHIALIS—*Arises*, tendinous and fleshy, from the middle part of the apex of the coracoid process of the scapula. Its fibres, as it descends, also arise from the edge of the short tendon of the biceps flexor cubiti. It forms a flat fleshy belly which is always perforated by the nerve, named Musculo-Cutaneous Externus.

Inserted, tendinous and fleshy, about the middle of the internal part of the os humeri, into a rough ridge.

Situation. This muscle is much connected with the

short head of the biceps flexor cubiti. In the arm it lies behind, and on the inside of the biceps, and is concealed partly by the pectoralis major and deltoides. It is inserted immediately below the tendons of the latissimus dorsi and teres major, and before the brachialis externus. The AXILLARY ARTERY is on its inner edge and the median nerve.

The internal intermuscular ligament (DEEP FASCIA of the ARM) is seen extending from the lower part of this muscle along a ridge to the internal condyle, and separating the brachialis internus from the brachialis externus, or third head of the triceps extensor cubiti.

Use. To move the arm upward and forward.

9. The BRACHIALIS INTERNUS—*Arises* from the middle of the os humeri, by two fleshy slips, which pass on each side of the insertion of the deltoid muscle; fleshy from all the forepart of the bone below, nearly as far as the condyles. The fibres converge, pass over the elbow-joint, and adhere to the capsular ligament.

Inserted, by a strong short tendon, into the rough surface immediately below the coronoid process of the ulna.

Situation. The belly is almost entirely concealed by the biceps flexor cubiti, excepting a small portion which projects beyond the outer edge of that muscle. The tendon dips down betwixt the supinator radii longus and pronator teres, crosses under the tendon of the biceps flexor, and is inserted on the inside of that tendon.

Use. To bend the forearm.

10. The TRICEPS EXTENSOR CUBITI is the great muscle which covers all the back part of the arm. It *arises* by three heads. The first, or long head, *arises,* by a broad tendon, from the inferior costa of the scapula near its cervix, and forms a large belly, which covers the back part of the os humeri. The second, or short head, *arises,* on the outer and back part of the os humeri, by an acute tendinous and fleshy beginning, from a ridge which runs from the back part of the great tuberosity toward the outer condyle. They also arise from the surface of bone behind the ridge, and from the inter-

muscular ligament which separates them from the muscles on the forepart of the arm.   The third head, called
BRACHIALIS EXTERNUS, *arises*, by an acute beginning,
from the inside of the os humeri above its middle, and
from a ridge extending to the inner condyle, from the
surface behind this ridge, and from the internal intermuscular ligament.

The three heads unite above the middle of the os
humeri, and invest the whole back part of the bone.
They form a thick strong tendon, which is

Inserted into the rough back part of the process of the
ulna, called Olecranon, and partly into the condyles of
the os humeri, adhering firmly to the capsular ligament.

Between the first and third head passes the musculo-
spiral nerve.   The ULNAR NERVE rests upon the front
part of the triceps muscle, accompanied by the INFERIOR
PROFUNDA artery from the brachial.   The SUPERIOR PRO-
FUNDA ARTERY accompanies the musculo-spiral nerve.

Situation.   The long head, where it arises from the
scapula, is concealed by the deltoid; it arises betwixt the
teres minor and teres major.   The short head arises immediately below the insertion of the teres minor.   The
tendon of the triceps sends off a thin fascia, which covers
the triangular surface of the ulna, on which we commonly
lean.   Numerous fibres are also sent off, to assist in forming the fascia of the forearm.

Use.   To extend the forearm.   The long head will also
assist in drawing the arm backward.

Dissection of the Fascia and Muscles situated on the Cubit or Forearm.

On removing the integuments of the forearm, we find
a strong fascia investing all the muscles.   It is attached
to the condyles, and it adheres firmly to the olecranon of
the ulna.   It receives, on the posterior part, a great addition of fibres from the tendon of the triceps extensor;
and on the forepart of the arm, it appears to be a continuation of the aponeurosis which is sent off from the
biceps flexor cubiti.

Above the fascia we meet with several cutaneous veins and nerves. The veins have been already described; the nerves are twigs of the branches of the brachial plexus, principally INTERNAL and EXTERNAL cutaneous.

The relative situation of the vessels at the bend of the arm should be well attended to. The cutaneous veins situated here vary much in size. The *vena basilica* is seen running over the forepart of the bend of the arm near the inner condyle, the *vena cephalica* situated near the outer condyle; and each of these veins receives a branch passing obliquely from the *vena mediana*. These vessels lie above the fascia, while the brachial artery lies just beneath the fascia, in a hollow resembling that of the axilla. It descends over the joint near the inner condyle, on the inside of the tendon of the biceps flexor cubiti, and under the aponeurosis sent off from that muscle to the common fascia of the forearm. It lies imbedded in cellular substance, betwixt the pronator teres and flexor muscles of the wrist and fingers on one side, and the supinator longus and extensor muscles on the other. In this hollow it divides into the radial ulnar and interosseal arteries. The artery is accompanied by two veins, and on its inner side runs the *radial nerve*.

Muscles situated on the Forepart of the Cubit, and arising from the Inner Condyle of the Os Humeri.

These are eight in number, and may be divided into two classes, the superficial and the deep seated.

First, the superficial.

All the muscles passing from the inner condyle may be said to arise by one common tendinous head from the condyle, and this head may be said to divide into the different muscles; but they will be here described as arising distinct from the condyle. It must, however, be recollected that their origins are intimately connected by intermuscular fascia, and that they cannot be separated without dividing some of their fibres.

1. The PRONATOR RADII TERES—*Arises*, tendinous and fleshy, from the anterior surface of the inner con-

dyle of the os humeri, and from the coronoid process of the ulna. It also arises from the fascia of the forearm. The fibres pass outward, run by the side of the tubercle of the radius, and pass over the outer edge of that bone, to be

Inserted, tendinous and fleshy, into a rough surface on the back part of the radius about its middle.

Situation. Of the muscles which pass from the internal condyle, the pronator teres is situated nearest the outer edge of the arm. Its tendon, to arrive at its place of insertion, passes under the belly of the supinator longus.

Use. To roll the radius, together with the hand, inward.

2. The FLEXOR CARPI RADIALIS—*Arises,* by a narrow tendinous beginning, from the lower and forepart of the internal condyle of the os humeri; fleshy from the fascia and intermuscular ligaments, and from the upper end of the ulna. It forms a thick belly, which runs down the forearm, and terminates in a flat tendon. This tendon passes under the annular ligament[1] of the wrist, runs through a groove in the os trapezium, and is

Inserted into the forepart of the base of the metacarpal bone of the forefinger.

Situation. This muscle is situated immediately under the fascia, excepting its upper extremity, over which the pronator teres crosses. Its insertion cannot be seen till the palm of the hand is dissected, where it will be found concealed by the muscles of the ball of the thumb.

Use. To bend the hand and to assist in its pronation.

3. The PALMARIS LONGUS—*Arises,* by a slender tendon, from the forepart of the inner condyle of the os

---

[1] The annular ligament of the wrist consists of two parts: 1. The ligamentum carpi transversale externum passes from the styloid process of the ulna and os pisiforme, transversely, over the back of the wrist, and spreads out broad, to be affixed to the styloid process of the radius. Under it pass the tendons of the extensor muscles. 2. The ligamentum carpi transversale internum is a strong ligament, which passes across the forepart of the wrist. It arises from the os pisiforme and os unciforme on the inner edge of the wrist, and is attached to the os scaphoides and os trapezius on the outer edge. Under it pass the tendons of the flexor muscles.

humeri, and fleshy from the intermuscular ligament; it forms a short fleshy belly, which soon sends off a long slender tendon. This tendon descends along the forearm, and is

Fig. 36.

THE MUSCLES OF THE FRONT OF THE FOREARM.

1. Lower part of the Biceps.
2. Part of the Brachialis Internus.
3. Edge of the Triceps.
4. Pronator Radii Teres.
5. Flexor Carpi Radialis.
6. Palmaris Longus.
7. One of the Divisions of the Flexor Sublimis Digitorum.
8. Flexor Carpi Ulnaris.
9. Palmar Fascia.
10. Palmar Brevis Muscle.
11. Abductor Pollicis Manus.
12. Portion of the Flexor Brevis Pollicis Manus.
13. Supinator Radii Longus.
14. Extensor Ossis Metacarpi Pollicis curving around the lower Border of the Forearm.

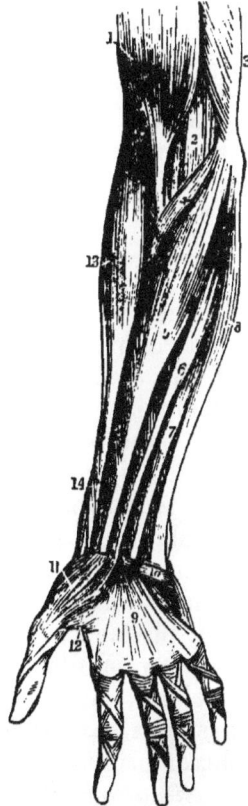

Inserted, near the root of the thumb, into the ligamentum carpi transversale internum, and into a tendinous membrane that covers the palm of the hand named Palmar Fascia, or APONEUROSIS PALMARIS.

Situation. It arises betwixt the flexor carpi radialis and flexor ulnaris. This muscle is sometimes wanting.

Use. To bend the hand, and stretch the palmar aponeurosis.

4. The FLEXOR CARPI ULNARIS—*Arises*, tendinous,

6*

from the inferior part of the internal condyle of the os humeri; tendinous and fleshy, from the inner side of the olecranon, and by a tendinous expansion from the posterior ridge of the ulna, to near the lower end of the bone. It also arises from the intermuscular fascia and fascia of the forearm. The fibres pass obliquely forward into a tendon which runs over the forepart of the ulna, and is

Inserted into the os pisiforme, and sometimes sends its fibres over a small ligament which goes to the base of the metacarpal bone of the little finger.

Situation. This muscle runs along the inner edge of the forearm, between the flexor sublimis on the forepart, and the extensor carpi ulnaris on the back part of the ulna.

Use. To bend the hand.

5. The FLEXOR SUBLIMIS PERFORATUS—*Arises*, tendinous and fleshy, from the under part of the internal condyle of the os humeri; tendinous, from the lower part of the coronoid process of the ulna; fleshy, from the tubercle of the radius, from the middle of the forepart of that bone, and from the middle third of its outer edge. These origins form a strong fleshy mass, which sends off four tendons. These tendons are connected by cellular membrane, and pass together under the annular ligament of the wrist; after which they separate, become thinner and flatter, pass along the metacarpal bone and first phalanx of each of the fingers, and are

Inserted into the anterior and upper part of the second phalanx, each tendon being, near the extremity of the first phalanx, divided for the passage of a tendon of the flexor profundus.

Situation. To expose the origin of this muscle, the bellies of the pronator teres, flexor carpi radialis, and palmaris longus, must be detached from the condyle. It descends along the forearm under these muscles, but a part of it is seen projecting toward the inner edge of the arm, betwixt the tendons of the palmaris longus and flexor carpi ulnaris. It arises from the radius below the insertion of the biceps flexor cubiti. Its tendons will be seen in the dissection of the palm of the hand.

ARTERIES OF THE FOREARM. 115

Use. To bend the second joint or phalanx of the fingers.

Notice the following bloodvessels and nerves, which will be discussed in detail, when the dissection of the arm has been finished.

Fig. 37.

A VIEW OF THE ARTERIES OF THE FOREARM.

1. The Lower Part of the Biceps Muscle.
2. The Inner Condyle of the Humerus, with the Humeral Origin of the Pronator Radii Teres and Flexor Carpi Radialis Muscles cut across.
3. The deep portion of the Pronator Teres Muscle.
4. The Supinator Longus Muscle.
5. The Flexor Longus Pollicis.
6. The Pronator Quadratus.
7. The Flexor Digitorum Profundus.
8. The Flexor Carpi Ulnaris.
9. The Anterior Annular Ligament. The figure is placed on the Tendon of the Palmaris Longus Muscle, divided close to its insertion.
10. The Brachial Artery.
11. The Great Anastomotic Artery.
12. The Radial Artery.
13. The Radial Recurrent Artery.
14. The Superficialis Volæ Artery.
15. The Ulnar Artery.
16. Its Superficial Palmar Arch giving Digital Branches to three fingers and a half.
17. The Great Artery of the Thumb (Magna Pollicis).
18. The Posterior Ulnar Recurrent.
19. The Anterior Interosseous Artery.
20. The Posterior Interosseous, as it is passing through the Interosseous ligament.

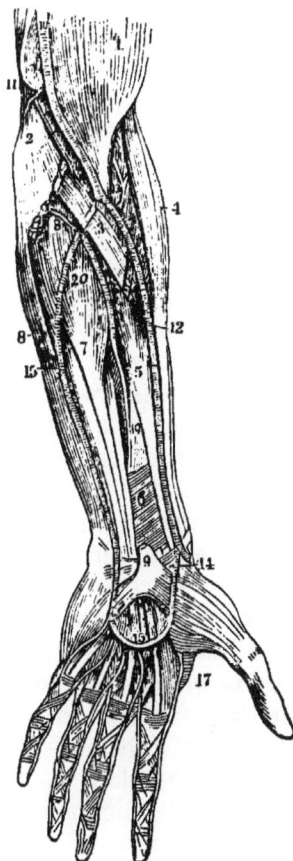

Brachial Artery at the bend of the arm, the median nerve on its inner side. The artery divides into RADIAL and ULNAR.

The RADIAL crosses to the radial side of the arm, at

the root of the thumb passes under its extensor tendons.
Its branches are

RECURRENS RADIALIS, toward external condyle.

MUSCULAR, to the muscles.

SUPERFICIALIS VOLÆ, to the ball of the thumb.

CARPAL. ANTERIOR and POSTERIOR. These are all
that can be seen at the present stage of the dissection.

RADIAL NERVE, to the outside of the artery.

A. ULNARIS.—Its first part deep under several muscles which arise from the internal condyle, and not visible,
emerges on the ulnar side of arm, and passes down the
arm between the FLEXOR CARPI ULNARIS and FLEXOR
SUBLIMIS DIGITORUM, over the annular ligament into
the hand, to form the arcus sublimis. The ulnar nerve
is to its ulnar or inner border. Its branches are

RECURRENS ULNARIS—Passes back toward the internal condyle.

INTEROSSEA—Too deep to be seen yet.

DORSALIS MANUS—Leaves the ulnar at the lower part
of the arm, and passes to the back of the wrist.

By removing the belly of the flexor sublimis, we expose the deep-seated muscles.

6. The FLEXOR PROFUNDUS PERFORANS—*Arises*,
fleshy, from the smooth concavity on the inside of the
ulna, betwixt the coronoid process and the olecranon;
from the smooth flat surface of the ulna, betwixt its posterior and internal angles; from the under part of the
coronoid process, and from the forepart of the ulna below that process. It also arises from the inner half of
the interosseous ligament. This muscle forms a thick
mass, which descends along the forepart of the ulna, adhering to that bone as low as one-third of its length
from its inferior extremity, and terminates in sending
off four tendons. These tendons pass together under
the annular ligament of the wrist, run through the slits
in the tendons of the flexor sublimis, and are

Inserted into the fore and upper part of the third or
last phalanx of all the fingers.

Situation. This muscle is concealed by the flexor sublimis and flexor carpi ulnaris. Its tendons will be seen
in dissecting the hand.

Use. To bend the last joint of the fingers.

7. FLEXOR LONGUS POLLICIS MANUS—*Arises,* by an acute fleshy beginning, from the upper and forepart of the radius, immediately below its tubercle, fleshy from the outer edge and anterior surface of that bone as low as two inches above its inferior extremity, and from the outer part of the interosseous ligament. It has also generally a tendinous origin from the internal condyle of the os humeri. This origin forms a distinct fleshy slip, which is joined to the inner and upper part of the portion of the muscle arising from the radius. The fibres pass obliquely into a tendon on the anterior surface of the muscle. The tendon passes under the annular ligament of the wrist, runs between the two heads of the short flexor of the thumb, and between the two sesamoid bones, and is

Inserted into the base of the extreme phalanx of the thumb.

Situation. This muscle lies by the side of the flexor profundus; the portion which arises from the inner condyle passes over the belly of the flexor profundus, and under the flexor sublimis. Its tendon will be seen in dissecting the short muscles of the thumb.

Use. To bend the last joint of the thumb.

On separating the lower part of the two last described muscles, we expose a small square muscle, passing transversely just above the wrist.

8. The PRONATOR QUADRATUS—*Arises,* broad, tendinous, and fleshy, from the inner edge of the ulna, extending from the lower extremity of the bone two inches up its edge. The fibres run transversely, adhere to the interosseous ligament, and are

Inserted into the lower and anterior part of the radius.[1]

Situation. This muscle lies close to the bones, covered by the flexor longus pollicis and flexor digitorum sublimis.

---

[1] This muscle, if carefully examined, will be seen to consist of two sets of fibres, having different directions—first pointed out, I believe, by Dr. J. Rhea Barton, of this city.

Use. To turn the radius, together with the hand, inward.

Fig. 38.

A FRONT VIEW OF SOME OF THE MUSCLES OF THE FOREARM.

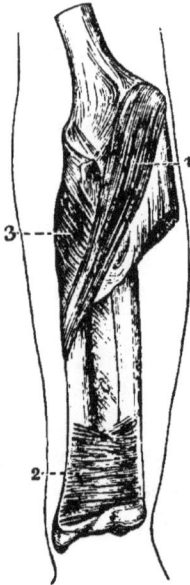

1. Pronator Radii Teres.
2. Pronator Quadratus.
3. Supinator Radii Brevis.

Muscles situated on the Outer and Back Part of the Forearm, and arising from the Outer Condyle of the Os Humeri.

These muscles are eleven in number, and may be divided into two classes : 1. The superficial; and 2. The deep seated.

The Superficial.

The muscles which arise from the outer condyle are much more distinct in their origins than those which arise from the inner condyle. Several of them arise a considerable way up the os humeri; but there is here also a common tendinous origin from which the extensor carpi radialis brevior, extensor digitorum communis, and extensor carpi radialis longior proceed, so that these muscles are intimately connected.

1. SUPINATOR RADII LONGUS—*Arises*, tendinous and fleshy, from the external ridge of the os humeri which leads to the outer condyle. It begins to arise nearly as far up as the middle of the bone, and ceases to adhere about two inches above the condyle. It forms a thick fleshy belly, which passes over the side of the elbow-joint, becomes smaller, and terminates above the middle of the forearm in a flat tendon. The tendon becomes gradually rounder, and is

Inserted into a rough surface on the outer side of the inferior extremity of the radius.

Situation. This muscle is situated immediately under the integuments along the outer edge of the arm and forearm. Its origin lies betwixt the brachialis internus and short head of the triceps extensor cubiti. Its insertion is crossed by the extensors of the thumb.

Use. To roll the radius outward, and turn the palm of the hand upward; also to bend the forearm on the humerus.

2. The EXTENSOR CARPI RADIALIS LONGIOR— *Arises*, tendinous and fleshy, from the external ridge of the os humeri, beginning immediately below the origin of the supinator longus, and continuing to arise as far as the upper part of the outer condyle. It forms a thick short belly, which passes over the side of the elbow-joint, and terminates above the middle of the radius in a flat tendon. The tendon runs along the radius, and, becoming rounder, passes through a groove in the back part of the inferior extremity of that bone, to be

Inserted into the posterior and upper part of the metacarpal bone of the forefinger.

Situation. The belly lies under the supinator longus, but part of it projects behind that muscle. The tendon descends behind that of the supinator, and passes under the extensors of the thumb and annular ligament of the wrist, to arrive at the place of its insertion.

Use. To extend the wrist and move the hand backward, and to assist in bending the forearm.

3. The EXTENSOR CARPI RADIALIS BREVIOR—
Arises, tendinous, from the under and back part of the
external lateral ligament of the elbow-joint. Its thick
belly runs along the outside of the radius, and terminates in a tendon, which passes through the same groove
in the radius as the extensor radialis longior, and under
the annular ligament.

Inserted, by a round tendon, into the upper and back
part of the metacarpal bone that supports the middle
finger.

Situation. This muscle lies partly under the extensor
radialis longior; but it also projects behind it. It passes
under the extensors of the thumb and the indicator.

Use. To extend the hand.

4. The EXTENSOR DIGITORUM COMMUNIS—*Arises*,
tendinous, from the under part of the external condyle
of the os humeri; fleshy, from the intermuscular fascia,
and from the inner surface of the fascia. It descends
along the back part of the forearm, and adheres to the
ulna where it passes over it. The fleshy belly terminates in four flat tendons, which pass under the annular
ligament in a depression on the back part of the radius,
and are

Inserted into the posterior part of all the bones of
the fingers by a tendinous expansion.

Situation. It arises and descends betwixt the extensor
radialis brevior and the extensor carpi ulnaris, and is
situated immediately under the integuments. The tendons are connected on the back of the metacarpal bone
by cross slips. The inner part of this muscle is sometimes described as a separate muscle, and is called EX-
TENSOR PROPRIUS MINIMI DIGITI, vel AURICULARIS.
It passes through a separate depression of the radius
and a particular ring of the annular ligament.

Use. To extend all the joints of the fingers.

The posterior surface of each finger is covered with a
tendinous expansion, which is formed by the tendons of
the common extensor, of the lumbricales, and interossei.
This tendinous expansion terminates in the third or extreme phalanx.

5. The EXTENSOR CARPI ULNARIS—*Arises*, tendinous, from the upper part of the external condyle; fleshy, from the intermuscular fascia and inside of the fascia.

Fig. 39.

THE SUPERFICIAL LAYER OF MUSCLES ON THE BACK AND FOREARM.

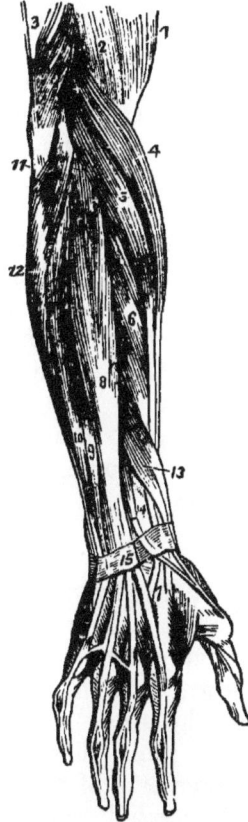

1. The Lower Part of the Biceps.
2. Part of the Brachialis Internus.
3. The Insertion of the Triceps into the Olecranon.
4. The Supinator Radii Longus.
5. The Extensor Carpi Radialis Longior.
6. The Extensor Carpi Radialis Brevior.
7. The Tendinous Insertion of these two Muscles
8. The Extensor Communis Digitorum.
9. The Extensor Minimi Digiti.
10. The Extensor Carpi Ulnaris.
11. The Anconeus.
12. Part of the Flexor Carpi Ulnaris.
13. The Extensor Minor Pollicis and the Ossis Metacarpi Pollicis lying together.
14. The Extensor Major Pollicis; its tendon is seen crossing the tendons of the two radio-carpal extensors.
15. The Posterior Annular Ligament. The tendons of the Extensor Communis are seen upon the back of the hand, and also their mode of distribution on the backs of the fingers.

It crosses toward the ulna, and arises, fleshy, from the back part of that bone. It terminates in a strong tendon, which passes through a groove in the back part of the lower end of the ulna, under the annular ligament, and is

Inserted into the posterior and upper part of the metacarpal bone of the little finger.

Situation. This muscle is entirely superficial. It

arises from the condyle betwixt the extensor digitorum communis and anconeus.

Use. To extend the wrist, and bring the hand backward; but chiefly to bend the hand laterally toward the ulna, as it will appear by pulling its tendon in the dissected subject.

6. The ANCONEUS is a small triangular muscle, situated at the outer side of the olecranon, immediately under the integuments.

It *arises*, tendinous, from the posterior and lower part of the external condyle of the os humeri; forms a thick triangular fleshy mass, adhering to the capsular ligament of the elbow-joint, and is

Inserted into the concave surface on the outside of the olecranon, and into the posterior edge of the ulna.

Situation. This muscle lies betwixt the upper part of the extensor carpi ulnaris and the olecranon. It is partly covered by the tendon of the triceps extensor cubiti, and is enveloped in a fascia sent off from that tendon.

Use. To assist in extending the forearm.

By removing the superficial muscles, we expose

The Deep Seated.

7. The SUPINATOR RADII BREVIS—*Arises*, tendinous, from the lower part of the external condyle of the os humeri; tendinous and fleshy, from the ridge running down from the coronoid process along the outer surface of the ulna. The fibres adhere firmly to the ligament that joins these two bones, pass outward round the upper part of the radius, and are

Inserted into the upper and outer edge of the tubercle of the radius, and into an oblique ridge extending from the tubercle downward and outward to the insertion of the pronator teres.

Situation. This muscle nearly surrounds the upper and outer part of the radius. It is concealed at the outer edge of the arm by the supinator longus and extensores carpi radiales; behind, by the extensor digitorum communis, extensor carpi ulnaris, and anconeus; before, by the brachialis internus, and by the tendon of the biceps flexor cubiti, close to which tendon this muscle is inserted.

Use. To roll the radius outward, and bring the hand supine.

On the back part of the forearm we meet with three muscles going to the thumb, and one to the forefinger.

8. The EXTENSOR OSSIS METACARPI POLLICIS— *Arises*, fleshy, from the middle and posterior part of the ulna, immediately below the termination of the anconeus, from the interosseous ligament, and from the posterior surface of the radius below the insertion of the supinator radii brevis. The fleshy fibres terminate in a tendon which passes through a groove in the outer edge of the lower extremity of the radius.

Inserted, generally by two tendons, into the os trapezium, and into the upper and back part of the metacarpal bone of the thumb.

Use. To extend the metacarpal bone of the thumb outwardly.

9. The EXTENSOR PRIMI INTERNODII POLLICIS MANUS—*Arises*, fleshy, from the back part of the ulna below its middle, from the interosseous ligament and radius. It runs along the lower edge of the extensor ossis metacarpi, and forms a tendon, which passes through the same groove as the tendon of that muscle, and is

Inserted into the posterior part of the first bone of the thumb. Part of the tendon is also continued into the base of the second or extreme phalanx.

Use. To extend the first phalanx of the thumb obliquely outward.

10. The EXTENSOR SECUNDI INTERNODII POLLICIS MANUS—*Arises*, tendinous and fleshy, from the posterior surface of the ulna above its middle, and from the interosseous ligament. Its belly partly covers the origins of the two other extensors of the thumb, and terminates in a tendon, which runs through a distinct groove in the back part of the radius, and is

Inserted into the posterior and upper part of the second or extreme phalanx of the thumb.

Use. To extend the last joint of the thumb obliquely backward.

Situation of the extensors of the thumb.—The origins of these muscles are concealed by the extensor digitorum communis and extensor carpi ulnaris. The tendon of the extensor secundi internodii is at a considerable distance from the tendons of the two other extensors; so that, in the intermediate space, we see the terminations of the tendons of the extensores carpi radiales. They invest the back part of the thumb with a fascia.

11. The INDICATOR—*Arises,* by an acute fleshy beginning, from the middle of the back part of the ulna, and from the interosseous ligament. Its tendon passes through the same sheath of the annular ligament with the extensor digitorum communis, and is

Inserted into the posterior part of the forefinger with the tendon of the common extensor.

Situation. It arises nearer to the inner edge of the arm than the extensor secundi internodii pollicis. It is concealed by the extensor digitorum communis and extensor carpi ulnaris. The tendon passes under that of the common extensor.

Use. To assist in extending the forefinger.

POSTERIOR INTEROSSEA ARTERY, resting on the posterior part of the interosseous ligament, supplying muscles, and anastomosing with carpal arteries on the back of the wrist.

Dissection of the Palm of the Hand.

The tendons which pass over the bones of the carpus into the palm of the hand are firmly bound down by the annular ligament of the wrist. They are invested and connected by cellular membrane, which forms sheaths, and secretes synovia to facilitate their motions.

On removing the integuments from the palm of the hand we meet with a strong fascia. It arises from the tendon of the palmaris longus and from the annular ligament of the wrist, expands over all the palm of the hand, and is fixed to the roots of the fingers, splitting to transmit their tendons. These forks or splits are connected by transverse fibres. This is the Fascia or APONEUROSIS PALMARIS. It is triangular. Where it arises from the wrist it is narrow, and does not cover the base of the metacarpal bones of the little and forefinger. As it runs over the hand it becomes broader, and is fixed by a bifurcated extremity in the lower end of each of the metacarpal bones of the four fingers. The palmar fascia is strong and thick, and conceals and supports the muscles of the hand. Its deep surface is connected to the interosseous fascia by two membranous prolongations. It exerts great influence on deep-seated abscesses of this part.

There is a small thin cutaneous muscle situated between the wrist and the little finger.

The PALMARIS BREVIS—*Arises* from the annular ligament of the wrist, and from the inner edge of the fascia palmaris.

Inserted, by small scattered fibres, into the skin and fat which cover the short muscles of the little finger and inner edge of the hand.

Use. To assist in contracting the palm of the hand.

The FASCIA PALMARIS may now be removed. Under it will be seen the arcus sublimis and the digital nerves from the ulnar and median presently to be described, and the four tendons of the flexor sublimis perforatus. They are seen coming from beneath the annular ligament of the wrist, and diverging as they pass toward their respective fingers. Each tendon splits at the extremity of the first phalanx for the passage of the tendon of the flexor profundus perforatus, and inserted into the base of the second phalanx..

Under the flexor sublimis are the four tendons of the flexor profundus perforans, which pass through the slits in the tendons of the former, and are inserted into the bases of the third phalanges of the fingers.

The LUMBRICALES are four small muscles, which *arise*, tendinous and fleshy, from the outer side of the tendons of the flexor profundus perforans, soon after those tendons have passed the ligamentum carpi annulare. Each of these muscles has a small belly, which terminates in a tendon. The tendon runs along the outer edge of the finger, and is

Inserted into the tendinous expansion which covers the back part of the phalanges of the fingers about the middle of the first joint.

Use. To bend the first phalanges of the fingers, the flexor profundus being previously in action, to afford them a fixed point.

The short muscles of the thumb and forefinger are five in number.

1. The ABDUCTOR POLLICIS MANUS—*Arises*, by a broad tendinous and fleshy origin, from the anterior surface of the annular ligament of the wrist, and from the os naviculare and os trapezium.

Inserted, tendinous, into the outer side of the root of the first phalanx of the thumb, and into the tendinous membrane which covers the back part of all the phalanges.

Situation. This muscle is situated immediately under the integuments, and is the outermost portion of the muscular mass forming the ball of the thumb.

Use. To draw the thumb from the fingers.

2. The FLEXOR OSSIS METACARPI POLLICIS, or Opponens Pollicis—*Arises*, broad and fleshy, from the annular ligament of the wrist, and from the os naviculare and os trapezium.

Inserted, tendinous and fleshy, into the anterior and lower part of the metacarpal bone of the thumb.

Situation. It lies under the abductor pollicis, and is almost entirely concealed; but a few of its fibres are seen projecting beyond the edge of that muscle.

Use. To bring the first bone of the thumb inward.

3. The FLEXOR BREVIS POLLICIS MANUS arises by two distinct heads.

(1) The outer head *arises* from the inside of the annular ligament; from the anterior surface of the os trapezium and os trapezoides, and from the root of the metacarpal bone of the forefinger.

Inserted into the outer sesamoid bone, which is connected by a ligament to the root of the first phalanx of the thumb.

Fig. 40.

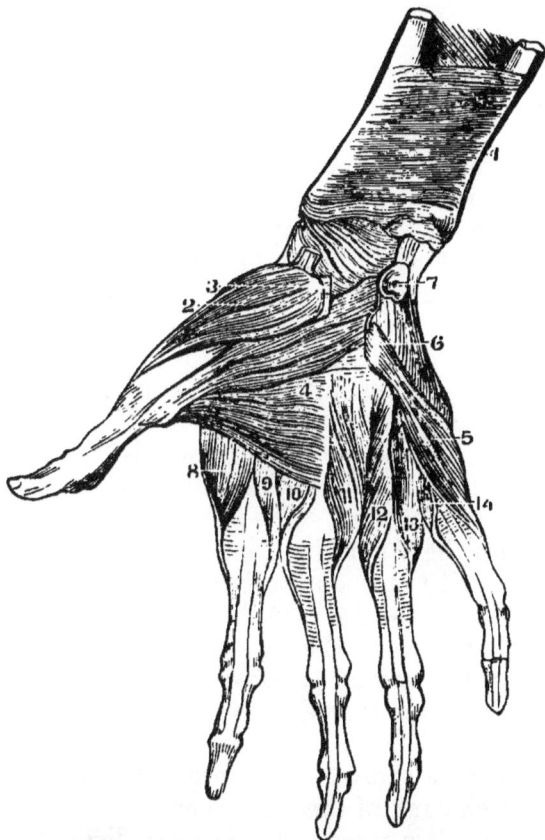

A FRONT VIEW OF THE DEEP-SEATED PALMAR MUSCLES.

1. Pronator Quadratus.
2. Opponens Pollicis.
3. Its attachment to the Annular Ligament.
4. Adductor Pollicis arising from the whole front of the second Metacarpal bone (Os Trapezium and Os Magnum).
5. Adductor Metacarpi Minimi Digiti.
6. Its Origin from the Os Unciforme.
7. Os Pisiforme.
8, 9, 10, 11, 12, 13, 14. Interossei Muscles.
8. Prior Indicis.
9. Posterior Indicis.
10. Prior Medii.
11. Posterior Medii.
12. Prior Annularis.
13. Posterior Annularis.
14. Interosseous Digiti Auricularis.

(2) The inner head *arises* from the upper part of the os magnum and os unciforme, and from the root of the metacarpal bone of the middle finger.

Inserted into the inner sesamoid bone, which is connected by a ligament to the root of the first phalanx of the thumb.

Situation. This muscle is in great part concealed by the abductor pollicis. Its inner origin is under the first umbricalis; its upper part is seen projecting, and between its two portions we find the tendon of the flexor longus pollicis.

Use. To bend the first joint of the thumb.

4. The ADDUCTOR POLLICIS MANUS—*Arises*, fleshy, from almost the whole length of the metacarpal bone sustaining the middle finger. The fibres converge, and pass over the metacarpal bone of the forefinger, to be

Inserted, tendinous, into the inner part of the root of the first phalanx of the thumb.

Situation. The belly of this muscle is concealed, as it lies close to the bone under the tendons of the flexor profundus and lumbricales. The tendon is seen where it is inserted into the thumb.

Use. To pull the thumb toward the fingers.

5. The ADDUCTOR INDICIS MANUS—*Arises*, tendinous and fleshy, from the os trapezium, and from the inner side of the metacarpal bone of the thumb. It forms a fleshy belly, runs over the side of the first joint of the forefinger, and is

Inserted, by a short tendon, into the outer side of the root of the phalanx of the forefinger.

Situation. This muscle is seen most distinctly on the back of the hand. It is there superficial, and is crossed by the tendon of the extensor secundi internodii pollicis. In the palm of the hand it is concealed by the muscles of the ball of the thumb.

Use. To move the forefinger toward the thumb, or the thumb toward the forefinger.

The insertion of the flexor carpi radialis is exposed by removing the muscles of the thumb.

The short muscles of the little finger are three in number.

1. The ABDUCTOR MINIMI DIGITI MANUS—*Arises*, fleshy, from the os pisiforme, and adjacent part of the annular ligament of the wrist. Its fibres extend along the metacarpal bone of the little finger.

Inserted, tendinous, into the inner side of the first phalanx, and into the tendinous expansion which covers the back part of the little finger.

Situation. The belly of this muscle is superficial. It is only covered by the straggling fibres of the palmaris brevis.

Use. To draw the little finger from the rest.

2. The FLEXOR PARVUS MINIMI DIGITI—*Arises*, fleshy, from the outer side of the os unciforme, and from the annular ligament of the wrist, where it is affixed to that bone.

Inserted, by a roundish tendon, into the base of the first phalanx of the little finger.

Situation. This muscle is also covered by the fibres of the palmaris brevis. It lies on the inner side of the abductor minimi digiti, and its tendon is connected to the tendon of that muscle.

Use. To bend the little finger, and bring it toward the other fingers.

3. ADDUCTOR METACARPI MINIMI DIGITI MANUS—*Arises*, fleshy, from the os unciforme and adjacent part of the annular ligament of the wrist. It forms a thick mass, which is

Inserted, tendinous, into the forepart of the metacarpal bone of the little finger, nearly its whole length.

Situation. It is concealed by the bellies of the abductor and flexor brevis minimi digiti.

Use. To bend and bring the metacarpal bone of the little finger toward the rest.

The INTEROSSEI are small muscles situated between the metacarpal bones, and extending from the bones of the carpus to the fingers. They are exposed by removing the other muscles of the thumb and fingers.

The INTEROSSEI INTERNI are seen in the palm of the hand, and are four in number. They arise, tendinous

7

and fleshy, from the base and sides of the metacarpal bones, and are inserted into the side of the first phalanx of the fingers, and into the tendinous expansion which covers the posterior surface of all the phalanges. The *Arcus profundus* runs across the metacarpal bones and these muscles.

1. The First, named Prior Indicis, *arises* from the outer part of the metacarpal bone of the forefinger; and is *inserted* into the outer side of the first phalanx of that finger. *Use.* To draw the forefinger toward the thumb.

2. The Second, named Posterior Indicis, *arises* from the root and inner side of the metacarpal bone of the forefinger; and is *inserted* into the inner side of the first phalanx of the forefinger. *Use.* To draw that finger outward.

3. The Third, named Prior Annularis, *arises* from the root and outer side of the metacarpal bone of the ring-finger; and is *inserted* into the outer side of the first phalanx of the same finger. *Use.* To pull the ring-finger toward the thumb.

4. The Fourth, named Interosseus Auricularis, *arises* from the root and outer side of the metacarpal bone of the little finger; and is *inserted* into the outer side of the first phalanx of the little finger. *Use.* To draw the little finger outward.

The internal interossei also assist in extending the fingers obliquely.

The INTEROSSEI EXTERNI, seu Bicipites, are three in number. They are larger than the internal, and are situated betwixt the metacarpal bones on the back of the hand. Each of these muscles *arise*, by a double head, from two metacarpal bones, and is inserted into the side of one of the fingers, and into the tendinous expansion which covers the posterior part of the phalanges.

1. The First, named Prior Medii, *arise* from the roots of the metacarpal bones of the fore and middle fingers; and is *inserted* into the outer side of the middle finger. *Use.* To draw the middle finger toward the thumb.

2. The Second, named Posterior Medii, *arises* from the roots of the metacarpal bones of the middle and ring-fingers; and is *inserted* into the inner side of the middle finger. *Use.* To draw the middle finger toward the ring-finger.

3. The Third, named Posterior Annularis, *arises* from the roots of the metacarpal bones of the ring and little fingers; and is *inserted* into the inner side of the ring-finger. *Use.* To draw the ring-finger inward. The external interossei also extend the fingers.

Of the Vessels and Nerves of the Superior Extremity.

Arteries.

The subclavian and axillary arteries have been described.

The BRACHIAL ARTERY may be said to have its course along the inside of the arm. Having left the axilla, it runs along the inferior edge of the coraco-brachialis. Rather higher up than the middle of the os humeri, it crosses over the tendinous insertion of that muscle, being here situated between the belly of the biceps flexor cubiti and the superior fibres of the brachialis externus. The artery then passes behind the inner edge of the biceps flexor cubiti, descending betwixt that muscle and the fibres of the brachialis internus. In dissecting this vessel, we find it invested by a fascia or sheath, formed by cellular membrane and some tendinous fibres. On dissecting this fascia, we find, close to the margin of the coraco-brachialis and biceps flexor cubiti, the *great median nerve;* under it the brachial artery, and, more superficially seated, the venæ comites and the vena basilica. As the artery approaches the lower extremity of the os humeri, it inclines forward toward the fold of the arm, and dives beneath the aponeurosis which arises from the inside of the tendon of the biceps flexor cubiti. Its situation at the fold of the arm has been described.

Branches of the Brachial Artery.

1. A. PROFUNDA HUMERI SUPERIOR, or Muscularis Superior, is sent off from the inner side of the brachial artery, immediately where it has left the fold of the arm-pit. It passes downward and backward round the os humeri, and is accompanied by the musculo-spiral nerve.

Fig. 41.

A VIEW OF THE AXILLARY AND BRACHIAL ARTERIES.

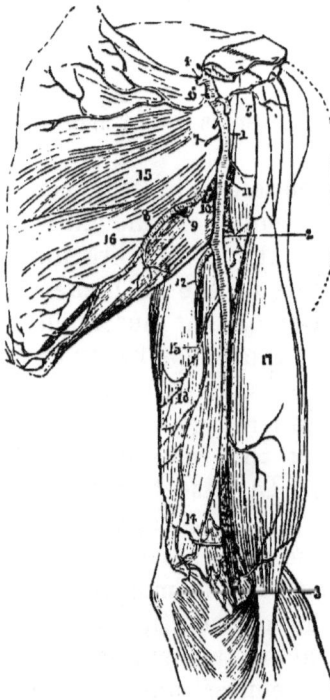

1. Axillary Artery, which ends at 2 in the Brachial.
2, 3. Brachial Artery.
4, 5, 6, 7. External Thoracic Arteries.
8. Subscapular Artery.
9. Its Dorsal Branch.
10. Posterior Circumflex.
11. Anterior Circumflex.
12. Profunda Superior.
13. Profunda Inferior vel Minor.
14. Anastomotic Artery.
15. Subscapularis Muscle.
16. Teres Major.
17. Biceps Flexor Cubiti.
18. Triceps.

It passes betwixt the brachialis externus and short head of the triceps extensor cubiti. Here it lies deep among the muscles, and divides into two branches. One accompanying the nerve spreads its ramification over the outer condyle, and anastomoses with the arteries below the elbow. The other branch is distributed along the inside of the arm and about the inner condyle.

2. A. PROFUNDA HUMERI INFERIOR is smaller than the last, and is sent off from the brachial artery about two inches lower down. It descends among the muscles on the inside of the arm, and is lost about the inner condyle.

3. The anastomosing or collateral arteries are as follows:

(1) The Ramus Anastomoticus Major passes from the inside of the brachial artery, about two or three inches above the inner condyle. It is distributed about the condyle, and its principal branch accompanies the ulnar nerve in the groove betwixt the olecranon and inner condyle, to anastomose with the recurrent branches of the arteries of the forearm.

(2) MUSCULAR BRANCHES.

The Brachial Artery, where it lies deep under the aponeurosis of the biceps, divides into three branches. 1. The radial. 2. The ulnar; and 3. The interosseous artery. The two last generally come off by one trunk, which subdivides.

1. ARTERIA RADIALIS. The radial artery is smaller than the ulnar, and in its course more superficial. It leaves the ulnar artery, and inclines toward the radial and outer edge of the forearm. At first it lies betwixt the pronator teres and supinator longus. It then descends close along the inner edge of the supinator longus, betwixt the supinator longus and flexor carpi radialis, and is accompanied by the radial nerve. Reaching the lower extremity of that bone, it divides into two branches.

(1) A. SUPERFICIALIS VOLÆ is by much the smallest of the two branches. It passes into the muscular mass which forms the ball of the thumb, anastomosing with the superficial palmar arch.

(2) The trunk of the radial artery crosses over the lower extremity of the radius to the back of the hand. It passes under the extensors of the thumb, and, arriving at the space betwixt the bases of the metacarpal

bones of the thumb and forefinger, plunges into the palm
of the hand.

Fig. 42.

ARTERIES OF THE FOREARM AND HAND.

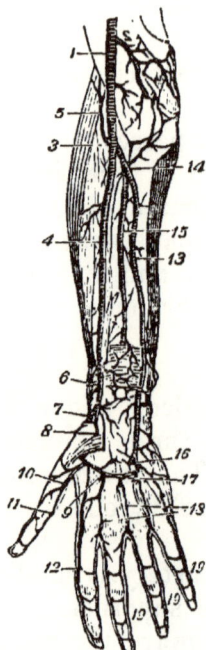

1. Brachial Artery.
2. Profunda Minor.
3. Bifurcation of the Brachial into the
   Radial and Ulnar.
4. Radial.
5. Recurrens Radialis.
6. Anterior Carpal.
7. Dorsalis Carpi.
8. Superficialis Volæ.
9. Arcus Profundus.
10. Magnus Pollicis.
11. Artery of the Thumb.
12. Radialis Indicis.
13. Ulnar Artery.
14. Recurrens Ulnaris.
15. Anterior Interosseal.
16. Cubitalis-manus Profunda, or Anasto-
    mosing Artery.
17. Arcus Sublimis.
18. Digital Arteries.
19, 19. Digito-ulnar Arteries.

The branches of the radial artery, in its course along
the forearm, are the following:

(1) The *recurrent artery* is sent off from the radial
immediately after it leaves the ulnar artery, and is distributed over the anterior part of the outer condyle,
where it anastomoses with branches of the brachial artery.

(2) MUSCULAR, to the muscles of the forearm.

(3) A branch leaves the artery immediately after it
has turned over the edge of the radius, and, ramifying
on the back of the hand, is named DORSALIS CARPI.

(4) Small vessels are sent off to the back part of the
thumb, named A. Dorsales Pollicis.

Having reached the palm of the hand, the radial artery divides into two branches.

(1) A. Pollicis, which sends two or three arteries along the anterior part of the thumb, and also often gives off a twig, the A. Radialis Indicis, which passes along the outer edge of the forefinger, and inosculates with a branch of the ulnar artery.

(2) The trunk of the radial artery forms the DEEP-SEATED PALMAR ARCH. From the root of the thumb, it passes across the metacarpal bones near their bases, and terminates at the metacarpal bone of the little finger, inosculating with a branch of the ulnar artery. This arch lies deep, close to the bones. It supplies the interosseous muscles and deep-seated parts of the palm, and some of its branches pass betwixt the metacarpal bones to the back of the hand.

2. ARTERIA ULNARIS. The ulnar artery is the largest branch of the brachial, and generally gives off the interosseous artery. It takes its course deep among the muscles on the inside of the forearm. It is seen passing under the pronator teres, flexor carpi radialis, palmaris longus, and flexor sublimis perforatus, but over the flexor profundus perforans. It descends in the connecting cellular membrane, between the flexor sublimis and profundus; but above the middle of the forearm, it emerges from these muscles, and appears at the ulnar edge of the arm, betwixt the flexor sublimis and flexor carpi ulnaris. It passes over the annular ligament of the wrist, but is covered by the fascia which ties down the tendon of the flexor carpi ulnaris. It passes under the palmar aponeurosis, on the inside of the os pisiforme, reaches the base of the metacarpal bone of the little finger, and begins to form the SUPERFICIAL PALMAR ARCH. This arch lies above the tendons of the flexor sublimis perforatus, immediately beneath the PALMAR APONEUROSIS. It crosses the metacarpal bones betwixt their bases and the middle of their bodies. It begins at the root of the little finger, and terminates at the root of the thumb, in inosculations with the branches of the radial artery. The convex side of the arch is turned toward the fingers, and sends off five branches.

(1) A branch to the muscles and inner edge of the little finger.

(2) Ramus digitalis primus, or the first digital artery, which runs along the space betwixt the two last metacarpal bones, and bifurcates into two branches, one to the outside of the little finger, and the other to the inner side of the ring-finger.

(3) The second digital artery, which bifurcates in a similar manner, and supplies the outer edge of the ring-finger, and the inner side of the middle finger.

(4) The third digital artery, which is distributed to the outer edge of the middle finger, and to the inner side of the forefinger.

(5) The ramus Pollicis ulnaris is the last branch of the ulnar artery, and is sent to the muscles of the thumb.

From the concavity of the arch are sent off the interosseous arteries of the palm, small twigs which supply the deep-seated parts, and perforate betwixt the metacarpal bones to the back of the hand.

The branches of the ulnar artery, in its course along the forearm and wrist, are the following:

(1) The RECURRENT ARTERIES are sent off from the ulnar artery immediately below the elbow. These arteries inosculate with branches of the brachial.

(2) Twigs to the muscles of the forearm.

(3) A. Dorsalis Carpi is sent off from the ulnar artery a little above the wrist to the back of the hand.

(4) A. PALMARIS PROFUNDA is sent off from the ulnar artery, on the inside of the os pisiforme. It passes into the flesh at the root of the little finger, and inosculates with the termination of the deep-seated palmar arch of the radial artery.

3. ARTERIA INTEROSSEA (or Interossea Communis). This artery is generally sent off from the ulnar. It immediately divides into two branches.

(1) The external or posterior interosseous artery is the smallest branch. It passes through the upper part of the interosseous ligament, to supply the muscles on the posterior part of the forearm. It sends off the A. Recurrens Interossea, which ramifies on the middle of the back part of the elbow-joint.

(2) The internal or anterior interosseous artery descends close upon the middle of the interosseous ligament, giving twigs to the adjacent muscles. At the upper edge of the pronator quadratus, it perforates the membrane to the back part of the arm, and spreads its extreme branches on the wrist and back of the hand.

Veins.

The cutaneous veins have been already described.

The brachial artery is accompanied by two veins, named Venæ Comites, or Satellites. These receive branches corresponding to the ramifications of the artery.

Nerves.

In the dissection of the axilla, we demonstrated the great axillary plexus, and traced its two first branches, the external scapular and circumflex nerves. See Fig. 20. The distribution of the five remaining branches of the plexus must now be described.

(3) The EXTERNAL CUTANEOUS NERVE (Musculocutaneus, or Perforans Casserii) is the third branch of the axillary plexus. It passes through the belly of the coraco-brachialis muscle. It continues its course betwixt the Biceps flexor cubiti and the Brachialis internus. It gives twigs to these muscles, and appears as a superficial nerve on the edge of the supinator longus. It runs over the outer condyle, and is distributed to the integuments on the outside of the forearm and back of the hand.

(4) The MEDIAN NERVE accompanies the brachial artery to the bend of the elbow. In its passage down the arm, it lies before that vessel, but at the elbow is situated on its inside. It gives off no branches until it has sunk under the aponeurotic expansion of the biceps flexor. Here it distributes many nerves to the muscles of the forearm, to the pronator teres, flexor carpi radialis, the flexors of the thumb and fingers, and the pronator quadratus. The trunk of the nerve perforates the pronator teres, passes betwixt the flexor digitorum sub-

limis and flexor profundus, and continues its course betwixt these muscles down to the wrist. Near the wrist it becomes more superficial, lying among the tendons of the flexors, and before it descends under the annular ligament, sends a superficial branch to the integuments and short muscles of the thumb. The nerve itself passes with the flexor tendons of the fingers under the annular ligament of the wrist, and appears on their outside, near the root of the thumb. It ramifies superficially in the hand, sending off four branches, which supply the thumb and all the fingers except the little finger and the ulnar side of the ring-finger.

(5) The ULNAR NERVE descends along the inside of the arm. It is at first situated immediately under the integuments, but below the middle of the arm is tied down by the intermuscular fascia. The nerve then runs between the inner condyle and the olecranon. After passing the condyle, it continues its course betwixt the two heads of the flexor carpi ulnaris, till it reaches the ulnar artery. It then accompanies the ulnar artery, lying on its inside, and running along the forearm betwixt the flexor ulnaris and flexor digitorum sublimis. It sends twigs to the neighboring muscles, and, when arrived near the wrist, divides into two branches. 1. The Smaller Branch, called Ramus posticus, passes under the tendon of the flexor carpi ulnaris, and over the lower end of the ulna, to be distributed to the back of the hand, and of the little and ring-fingers. 2. The continued trunk of the nerve passes, on the inside of the ulnar artery, over the annular ligament of the wrist. It passes under the palmar aponeurosis, and divides into three principal branches. The first supplies the integuments and muscles on the ulnar edge of the hand, and the inner side of the little finger. The second is distributed to the outer side of the little finger and inner side of the ring-finger. The third branch accompanies the deep-seated palmar arch, and terminates in the short muscles of the thumb and forefinger, communicating with the median nerve.

(6) The MUSCULO-SPIRAL NERVE (RADIAL of some

anatomists) passes from the axilla behind the os humeri,
making a spiral turn round the bone to reach the outside
of the arm.   It first descends between the brachialis
externus and short head of the Triceps extensor cubiti,
accompanying the arteria profunda humeri superior, and
passing deep into the flesh of the arm.   Before it makes
this turn, it gives branches to the muscles, also a cutaneous branch, which descends on the inside of the arm.
From the back part of the arm the great trunk of the
nerve is reflected spirally forward.   It is seen emerging
betwixt the supinator longus and brachialis internus,
seated deep and close to the bone.   It descends betwixt
these muscles, keeping close to the edge of the supinator
longus.   Immediately after passing the fold of the arm,
it sends off a nerve, which descends, superficial, upon the
radial edge of the forearm, as far as the wrist; and, at
this point, the trunk of the muscular nerve divides itself
into two branches.   The first, a large branch, perforates
the supinator brevis, and supplies the extensor muscles
of the hand and fingers.   The second branch accompanies the supinator longus down the forearm, and near
the wrist turns under the tendon of that muscle, over
the edge of the radius.   It then divides into several
branches which ramify on the back of the wrist, thumb,
and forefinger.

(7) The INTERNAL CUTANEOUS NERVE descends, superficial, along the inside of the arm and forearm.   It
was described among the cutaneous nerves of the arm.

CHAPTER X.

DISSECTION OF THE ABDOMEN.

Muscles of the Abdomen, and the Parts connected with them in Dissection.

BEFORE commencing its dissection, study what are called its regions. To mark them out, drop a perpendicular from the anterior inferior spinous processes through the cartilages of the ribs. Then cross these two lines by two others drawn the upper one through the points where the first two touched the cartilages of the ribs, and the second from one anterior superior

Fig. 43.

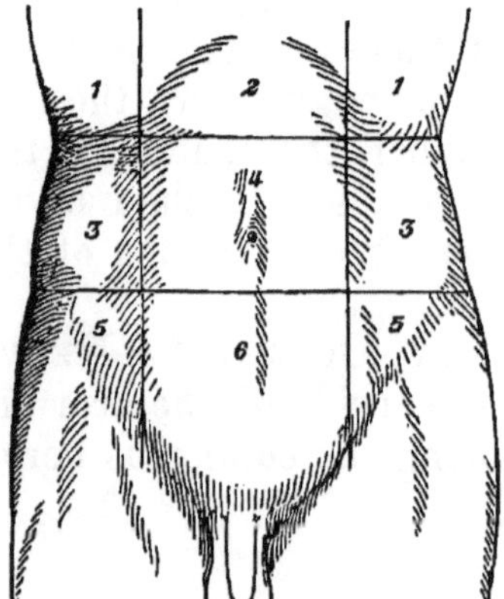

THE ABDOMINAL REGIONS.

1, 1. Hypochondriac Regions.   4. Umbilical Region.
2. Epigastric Region.   5, 5. Iliac Regions.
3, 3 Lumbar Regions   6. Hypogastric Region.

spinous process of the ilium to the other. These lines will define nine regions. Three in the centre, which, enumerated from above downward, are the EPIGASTRIC, UMBILICAL, and HYPOGASTRIC. Three on either side, which from above downward are the HYPOCHONDRIAC, LUMBAR, and ILIAC REGIONS.

The muscles are ten in number, five on each side.

An incision should be made through the integuments, from the sternum to the os pubis; and this should be crossed by another passing from the lower end of the sternum obliquely upward toward the axilla; dissect off the layers in order and this will expose—

In the superficial fascia of the abdomen an artery, accompanied by its veins, the SUPERFICIAL EPIGASTRIC, or Arteria ad cutem abdominis.

1. The OBLIQUUS DESCENDENS EXTERNUS.—*Origin.* By eight triangular fleshy slips from the lower edges and external surfaces of the eight inferior ribs, at a little distance from their cartilages; the five superior slips meet on the ribs an equal number of the digitations of the serratus major anticus, and the three inferior are connected with the attachments which the latissimus dorsi has to the rib. To gain a complete view of this muscle, the neighboring portions of the pectoralis major, serratus anticus, and latissimus dorsi should be dissected with it.

The muscular fibres proceed obliquely downward and forward, and about the middle of the side of the belly terminate abruptly in a thin broad tendon, which is continued in the same direction over all the forepart of the belly. Here it covers the anterior surface of the rectus abdominis; it is very thin at the upper part, where the rectus lies on the cartilages of the ribs, and is often removed by the beginner, unless he is very cautious.

Insertion. Tendinous and fleshy, into two anterior thirds of the outer edge of the crista of the os ilium, from the anterior superior spine of which, for it extends to the os pubis, forming Poupart's ligament, into the ensiform cartilage, and into the whole length of the linea alba.

Use. To draw down the ribs in expiration, to bend the trunk forward when both muscles act, or to bend it obliquely to one side when one of them acts singly; to raise the pelvis obliquely when the ribs are fixed; to compress the abdominal viscera, to thrust the diaphragm upward, to assist in the expulsion of the urine and feces, and of the fœtus.

In the course of the dissection of this single muscle, the following points must be attended to:

The LINEA ALBA, a white line running along the middle of the abdomen, from the cartilago ensiformis to the os pubis; formed by the tendinous fibres of the two obliqui and the transversalis muscles, interlaced with those of the same muscles on the opposite side; it is half an inch broad at the navel; and decreases gradually both above and below that part; but particularly in the latter situation, where it is reduced at last to a mere line.

LINEA SEMILUNARIS, a semicircular white line running from the os pubis obliquely upward over the side of the abdomen, at the distance of about four inches from the linea alba; formed by the tendons of the two oblique and transverse muscles uniting at the edge of the rectus, before they separate to form the sheath for that muscle.

LINEÆ TRANSVERSÆ, three or four white lines, crossing from the linea semilunaris to the linea alba; formed by the tendinous intersections of the recti shining through the strong sheath which covers them. These are not evident in all subjects in this stage of the dissection.

UMBILICUS, or Navel. This which, before the integuments were removed, was a depression, appears now a prominence; it consists of condensed cellular membrane.

ANNULUS ABDOMINALIS, or Ring, an oblique slit or opening just above the angle of the pubis; formed by the tendon of the external oblique, divided into two portions called the *pillars* or *columns* of the ring, of which one (the *superior* or *internal*) is attached to the symphysis, and the other (the *inferior* or *external*) to the tuberosity of the pubis; and allowing a passage to the spermatic cord in the male, and the ligamentum teres of

Fig. 44.

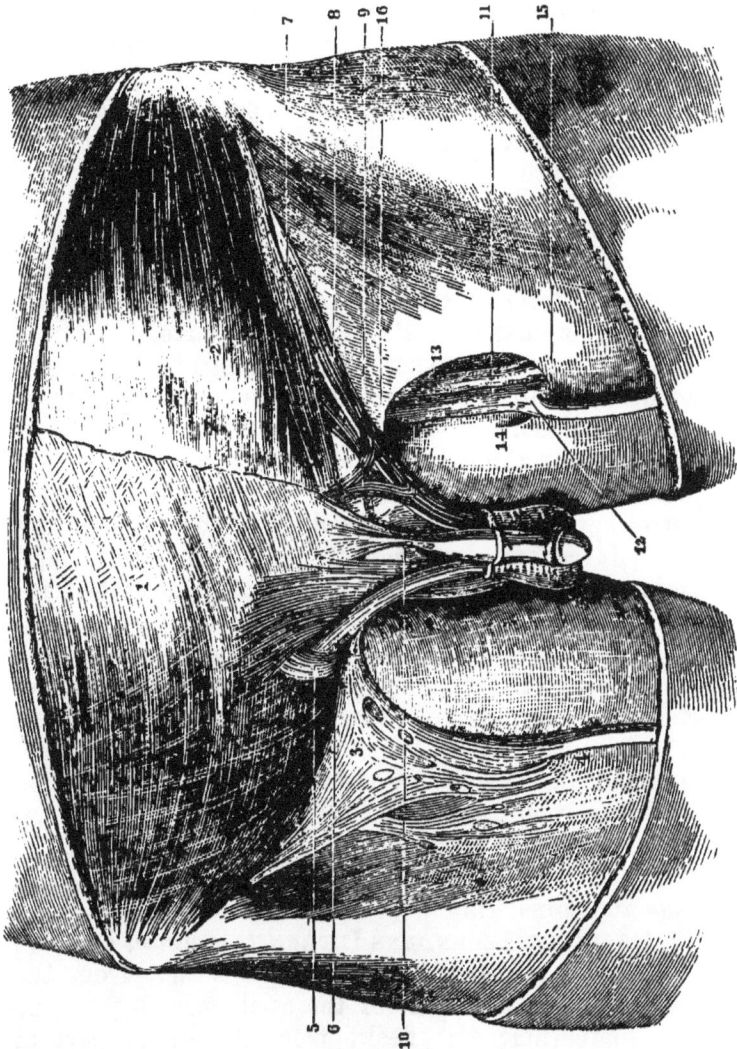

DISSECTION OF SOME OF THE PARTS CONCERNED IN FEMORAL AND INGUINAL HERNIA.

1. Tendon of the External Oblique Muscle.
2. Tendon of the Internal Oblique, the first-named muscle being dissected off.
3. Cribriform Fascia.
4. Vena Saphena.
5. External Abdominal Ring and Spermatic Cord.
6. Poupart's Ligament.

the uterus in the female. This slit is triangular; the pubis is the base, the two columns are the two sides of the triangle. From the margins of the pillars a thin fascia is derived which passes down upon the cord (the INTERCOLUMNAR or SPERMATIC FASCIA). It forms a coating for an inguinal hernia. The apex has a rounded figure in consequence of some transverse fibres, which connect the two columns where they first separate; and it points obliquely upward and outward. It belongs to the external oblique alone, there being no such opening, either in the internal oblique, or the transversalis; it is much smaller in the female than in the male.

LIGAMENTUM POUPARTII, a strong ligament, stretching from the anterior superior spinous process of the os ilium, to be fixed to the spine of the os pubis. This in truth is merely the lower edge of the tendon of the obliquus externus abdominis, extended from the anterior superior spinous process of the ilium to the angle of the pubis. It covers the femoral vessels and nerves, and certain muscles, and has lately been often described under the name of the CRURAL ARCH.

Dissect off the serrated origin of the external oblique from the ribs and from the spine of the os ilium, and detach it from the obliquus internus, which lies below it, and which is connected to it by loose cellular substance, and by small vessels. Continue to separate the two muscles till you find their tendons firmly attached, *i.e.* a little way beyond the linea semilunaris. Separate the tendon from Poupart's ligament to within half an inch of the abdominal ring.

2. OBLIQUUS ASCENDENS INTERNUS—*Arises* by short tendinous fibres, which soon become fleshy, from the whole length of the spine of the os ilium, and from the fascia lumborum; also fleshy from the upper part of Poupart's ligament at the part next to the os ilium.

The fibres run in a radiated direction; those which originate from the back part of the os ilium run obliquely upward; those from the forepart of the ilium pass more transversely; and from Poupart's ligament the fibres descend. The fleshy belly is continued rather

Fig 45.

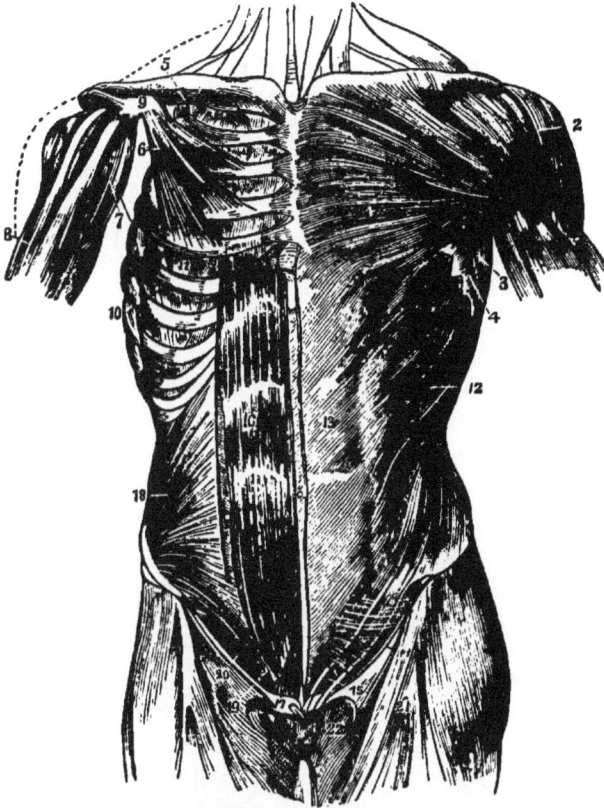

THE MUSCLES OF THE ANTERIOR ASPECT OF THE TRUNK; ON THE LEFT SIDE THE SUPERFICIAL LAYER IS SEEN, AND ON THE RIGHT THE DEEPER LAYER.

1. The Great Pectoral Muscle.
2. The Deltoid Muscle.
3. The Anterior Border of the Latissimus Muscle.
4. The Indigitations of the Great Serratus Muscle.
5. The Right Subclavian Muscle.
6. The Small Pectoral Muscle.
7. The Coraco-Brachialis Muscle.
8. The Upper Part of the Biceps Muscle, showing its two heads.
9. The Coracoid Process of the Scapula.
10. The Great Serratus Muscle of the Right Side.
11. The External Intercostal Muscle of the Fifth Intercostal Space.
12. The External Oblique Muscle.
13. Its tendon or Aponeurosis; on the left of this number is the semilunar line, and on the right the middle white line (linea alba).

14. Poupart's Ligament or the Crural Arch
15. The External Inguinal or Abdominal Ring; the crescentic open-
    ing to the right of 15 is the saphenous opening in the Femoral
    Aponeurosis.
16. The Rectus Abdominis Muscle of the Right Side brought into
    view by the removal of the anterior segment of the sheath
    formed by the tendons of the Broad Muscles of the Abdomen.
17 The Pyramidal Muscle.
18. The Internal Oblique Muscle.
19. The Conjoined Tendon of the Internal Oblique and Transver-
    salis Muscle
20. The Arch formed by the Lower Border of the Internal Oblique
    and Transversalis Muscles, from beneath which the Spermatic
    Cord has been removed.
21. Fascia Lata Femoris.
22. Saphenous Opening.
The Crescentic Edge of the Sartorial Fascia is seen just above Fig.
    22, and the Interior or Pubic Point of the Crescent is known as
    Hey's Ligament.

more forward than that of the external oblique before it
terminates in a flat tendon.

Inserted into the cartilages of the six or seven lower
ribs; fleshy into the three inferior, and, by a tendinous
expansion, which is extremely thin, resembling cellular
membrane, into the four superior, and also into the ensiform cartilage. The sheet of tendon in which the
fleshy belly ends is continued, single and undivided, into
the linea semilunaris, where, adhering pretty firmly to
the tendons of the obliquus externus and transversalis,
it divides into two layers. The anterior and more considerable layer joins the tendon of the external oblique,
and runs over the rectus to be inserted into the whole
length of the linea alba. The posterior and thinner
layer, adhering to the anterior surface of the transversalis, passes into the linea alba behind the rectus as low
as half way between the umbilicus and os pubis; but
below this place the whole tendon of the internal oblique
passes along with that of the external oblique before the
rectus, and is inserted into the lower part of the linea
alba. The inferior edge of the muscle extends in a
nearly straight direction over the spermatic cord to be
fixed by a tendinous attachment to the tuberosity of the
pubis.

Situation. It is covered by the obliquus descendens
externus and latissimus dorsi.

Use. To assist the obliquus externus; but it bends the trunk in the reverse direction, so that the muscle on each side co-operates with the obliquus ˚externus of the opposite side.

About the middle of Poupart's ligament, a delicate fasciculus of fibres is sent off from this muscle over the spermatic cord, where it passes under its edge in its way to the ring. This is named the

CREMASTER, and is continued down to the cord till it is insensibly lost on the tunica vaginalis testis. It will be seen in the dissection of the scrotum. It forms a covering for oblique inguinal hernia, as from what will be presently seen it overlies the inguinal canal. Its *use* is to suspend, draw up, and compress the testicle.

We must now dissect the attachments of the internal oblique from the cartilages of the ribs, from the fascia lumborum, and from the spine of the os ilium, and, by continuing our dissection from behind forward, separate it from the transversalis abdominis, which lies under it. This separation may be continued as far as where the tendons of the two muscles are inseparable, *i.e.* rather more forward than the linea semilunaris. As this muscle lies very close upon the transversalis, caution is required to avoid detaching both muscles together. Let the student begin his separation at the crista of the ilium, where the course of the CIRCUMFLEXA ILII ARTERY and VEIN will show him when he has arrived at the surface of the transversalis.

3. TRANSVERSALIS ABDOMINIS — *Arises*, tendinous, from the fascia lumborum and back part of the spine of the os ilium; fleshy from all the remaining part of the spine of the ilium and from the inner surface of Poupart's ligament; and fleshy from the inner·or back part of the cartilages of the seven lower ribs, where its fibres meet those of the diaphragm.

The fleshy fibres proceed transversely, and end in a flat sheet of tendon, which, after being connected to the other tendons at the linea semilunaris, passes with the posterior layer of the internal oblique behind the rectus, and is *inserted* into the ensiform cartilage, and into the whole length of the linea alba, excepting its lowermost part; for, at the middle distance between the umbilicus and os pubis, a slit or fissure is formed in this tendon, through which the rectus abdominis passes; and the remainder of the tendon passes before the rectus, to be inserted into the lower part of the linea alba. Its inferior edge is connected with that of the internal oblique muscle, and the two form a common tendon (the CONJOINED TENDON), which is inserted into the LINEA INNOMINATA, which would place it behind the external abdominal ring, and thus strengthen it.

Use. To support and compress the viscera of the abdomen.

When the transversalis is detached from its origins, and turned back toward the linea semilunaris, it will be seen to be lined by a fascia, strongest near to Poupart's ligament, to which it is attached. This is named by Sir Astley Cooper, its discoverer, the FASCIA TRANSVERSALIS, and prevents the bowels from being protruded under the inferior margins of the obliquus internus and transversalis muscles. It is perforated about the middle between the ilium and pubes, by an opening (INTERNAL ABDOMINAL RING) for the passage of the spermatic cord, which then goes obliquely downward, inward, and forward to the ring of the external oblique. The space between this internal ring in the fascia transversalis and the external ring in the tendon of the external oblique is called the INGUINAL CANAL, and is traversed by an oblique inguinal hernia. If the cord be drawn out, the fascia transversalis follows, being prolonged into a funnel-shaped figure. This prolongation forms the INFUNDIBULAR FASCIA of some, and is enumerated often as a covering of inguinal hernia.

In VENTRO-INGUINAL or DIRECT HERNIA, the intestine comes out only through the external ring instead of entering the canal at the internal ring; and, as the CONJOINED TENDON is inserted behind the external ring, it must carry this tendon before it (according to some), and will not receive a covering from the cremaster, as the cord which it overlies will be to the outer side of the tumor.

The sheath of the rectus is now to be attended to: it is formed by the tendons of the three other muscles, viz.: the two obliqui and the transversalis. These, when they reach the edge of the rectus, form the appearance named Linea Semilunaris; they then split and inclose the rectus in their duplicature; the whole tendon of the external oblique, with the anterior layer of the internal oblique, passes before the rectus; and the whole posterior layer of the internal oblique, together with the whole tendon of the transversalis muscles, pass behind the rectus, excepting at the lower part; but, for two or three inches above the pubis, all the tendons go in front of the muscle, and the posterior part of the sheath is consequently deficient, the rectus lying naked on the peritoneum, or having a very thin expansion of the transversalis fascia.

The two oblique muscles are now to be replaced; then, making an incision by the side of the linea alba, and thus opening the sheath of the rectus through its whole length, you dissect it back toward the linea semilunaris, and thus lay bare the fibres of the muscle next to be described.

4. RECTUS ABDOMINIS—*Arises*, by a flat tendon, from the forepart of the os pubis; as it ascends, its fleshy belly becomes broader and thinner.

Inserted, by a thin fleshy expansion, into the ensiform cartilage, and into the cartilages of the three inferior true ribs.

Situation. This pair of muscles is situated on each side of the linea alba, under the tendons of the oblique muscles. The muscle is generally divided by three tendinous intersections; the first is at the umbilicus, the second where it runs over the cartilage of the seventh rib, and the third in the middle between these; and there is commonly a half intersection below the umbilicus. By these intersections, the muscle is connected firmly to the interior part of its sheath, forming the LINEÆ TRANSVERSÆ, while it adheres very slightly by loose cellular substance to the posterior layer.

Use. To compress the forepart of the abdomen, to bend the trunk forward, or to raise the pelvis.

On each side of the linea alba, and inclosed in the lower part of the sheath of the rectus, is sometimes found a small muscle, named

5. PYRAMIDALIS.—*Origin*. Tendinous and fleshy, of the breadth of an inch from the os pubis, anterior to the origin of the rectus.

Insertion. By an acute termination, near half-way be tween the os pubis and umbilicus, into the linea alba and inner edge of the rectus muscle.

Use. To assist the lower part of the rectus.

Dissection of the Cavity of the Abdomen.

Make a longitudinal incision from the scrobiculus cordis to the umbilicus, and from that point an oblique incision on each side toward the anterior spinous process of the os ilium, forming thus three triangular flaps. In doing this, avoid cutting the intestines, by raising the muscles from them after the first puncture.

Before you disturb the viscera, observe the general situation of those parts which appear on the first opening of the abdomen.

1. The internal surface of the PERITONEUM, smooth, shining, and colorless, covering the parietes of the abdomen, and the surface of all the viscera.

2. In the triangular portion of integument folded down over the pubes, three ligamentous cords project through the peritoneum, two running laterally, and the other in the middle, toward the navel. These are the remains of the two umbilical arteries and the urachus.

3. The epigastric artery, accompanied by two veins, may be seen through the peritoneum, ascending obliquely upward and inward from under Poupart's ligament. Its origin from the external iliac artery, and its relation to the internal abdominal ring should be noted, being on its posterior wall; and, therefore, when a hernial stricture occurs here, the incision should be made upward, to avoid it. Close along Poupart's ligament will be seen the CIRCUMFLEX ILII artery, running toward the crest of the ilium. It comes from the external iliac artery.

4. The upper edge of the liver is seen extending from the right hypochondriac region, across the epigastric, into the left hypochondriac region; in it a fissure is seen, into which enters, inclosed in a duplicature of peritoneum, the ligamentum teres, which was, in the foetus, the umbilical vein. The fundus of the gall-bladder, if distended, is sometimes seen projecting from under the edge of the liver.

5. The STOMACH will be found lying in the left hypochondriac region, and upper part of the epigastric; but if distended, it protrudes into the umbilical region.

6. The GREAT OMENTUM proceeds from the great curvature of the stomach, and stretches down like a flap over the intestines.

7. The GREAT TRANSVERSE ARCH OF THE COLON will be seen projecting through the omentum; it mounts up from the iliac fossa of the right side, crosses the belly under the edge of the liver, and under the greater curvature of the stomach, and descending again upon the left side, sinks under the small intestines, and rests upon the wing of the left iliac fossa.

8. The SMALL INTESTINES lie convoluted in the lower part of the belly, surrounded by the arch of the colon.

Such is the general appearance on first opening the abdomen; but this will vary somewhat, as one intestine may happen to be more inflated than another, or as the position of the body may have been after death.

It will now be proper to consider the parts more minutely.

1. The PERITONEUM.—Observe how it is reflected from the parietes of the abdomen over all the viscera, so that they may be said to be situated behind it; trace its reflections from side to side, and from above downward; you will see that the external coat of every viscus, and all the connecting ligaments, are reflections or continuations of this membrane.

(1) The FOUR LIGAMENTS of the LIVER are formed by the peritoneum, continued from the diaphragm and parietes of the abdomen.

152 PRACTICAL ANATOMY.

Fig. 46.

REFLECTIONS OF THE PERITONEUM.

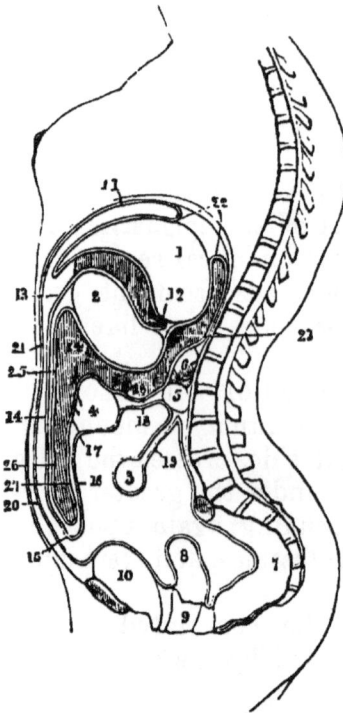

1. Liver.
2. Stomach.
3. Small Intestine.
4. Arch of the Colon.
5. Duodenum.
6 Pancreas.
7. Rectum.
8. Uterus.
9. Vagina.
10. Bladder.
11. Peritoneum reflected a little farther back, from the Diaphragm to the Liver, which last it covers above in front and below, and forms the Anterior Lamina of the Lesser Omentum.
12. It then covers the anterior face of the stomach, and forms at 13 and 14 the anterior layer of the omentum majus: at
15. It is reflected upward to form at 16 the posterior layer of that omentum; at
17. It embraces the colon on its posterior surface and forms the posterior lamina of the mesocolon at 18; it then passes in front of the duodenum, 5, and descends to embrace the small intestine, 3, whence it is reflected upward so as to give the posterior lamina to the mesentery, 19; it next passes down the posterior parietes of the abdomen, covers the rectum, 7, in front, the uterus, 8, the bladder, 10, and thence ascends to constitute the abdominal peritoneum, 20 and 21, lines the diaphragm, and terminates above in the coronary ligament of the liver at 22. If we now trace the peritoneum from the posterior margin of that ligament, 22, we find it coating the posterior face of the stomach, 1, and then separating from that organ to form the posterior lamina of the lesser omentum at 23; it next covers the posterior face of the stomach, 24, and is thence reflected downward to constitute the posterior layer of the anterior fold of the greater omentum, 25, 26; after which it turns upward, and forms at 27 the anterior layer of the posterior fold of the greater omentum; it then invests the front surface of the colon, 4, and forms at 28 the anterior face of the mesocolon; it thence passes upward in front of the pancreas, 6, and terminates where we began, at the posterior margin of the coronary ligament of the liver.

a. The MIDDLE or SUSPENSORY LIGAMENT, inclosing in its duplicate the LIGAMENTUM TERES.

b. The CORONARY LIGAMENT, connecting the upper surface of the Liver to the diaphragm.

c. The BROAD LIGAMENT of the right side.

d. The BROAD LIGAMENT of the left side.

(2) The LESSER OMENTUM, or HEPATICO-GASTRIC OMENTUM, is formed by two laminæ of peritoneum, passing from the under surface of the liver to the lesser curvature of the stomach, and containing in its duplicate the vessels of the liver.

(3) The GREAT EPIPLOON, GASTRO-COLIC or OMENTUM MAJUS.—Observe, that the peritoneum, coming from both surfaces of the stomach, and from the spleen, proceeds downward into the abdomen, and is then reflected back upon itself, till it reaches the transverse arch of the colon, where its laminæ separate to invest that intestine. This reflection is named the Great Omentum; it is a pouch or bag, composed of four laminæ of peritoneum, and the opening into it is by the FORAMEN OF WINSLOW: observe the situation of this semilunar opening; it is on the right side of the abdomen, at the root of the lesser lobe, or lobulus spigelii of the liver; it leads under the lesser omentum, under the posterior surface of the stomach, but above the pancreas and colon, into the sac of the omentum;—the omentum sometimes reaches to the lower part of the hypogastric region, sometimes not beyond the navel; it contains in its duplicature more or less of adipose substance.

(4) The MESENTERY.—Observe, that the peritoneum, reflected from each side of the vertebræ, proceeds forward, to connect the intestines loosely to the spine; that it begins opposite to the first lumbar vertebra, crosses obliquely from left to right, and ends half-way between the last lumbar vertebra and the groin. At its commencement, it binds down the extremity of the duodenum, and it terminates where the head of the colon begins. The great circumference which is in contact with the intestines is very much plaited or folded, and is several yards in length. Between the laminæ of the mes-

entery, observe the MESENTERIC GLANDS, the branches
of the superior mesenteric artery ramifying and forming
arches; the mesenteric veins accompanying the arteries;
the trunk of the lacteals, situated contiguous to the mesenteric artery on its left side. It may sometimes be inflated by the blowpipe. Nerves also run in the mesentery, but are not easily demonstrated.

(5) The MESOCOLON is similar to the mesentery, and
connects, in like manner, the colon to the spine.

2. HEPAR, the LIVER.—*Situation.* Partly in the right
hypochondrium, which it fills up, reaching as low as the
kidney of that side, partly in the epigastrium, and running also some way into the left hypochondrium.

Connected by its four ligaments to the inferior surface
of the diaphragm, and by the lesser epiploon to the small
curvature of the stomach. The little epiploon should now
be removed, to discover the different parts of the liver.

Observe the superior or convex surface adapted to the
arch of the diaphragm; the inferior or concave surface
resting on the stomach;—the posterior or thick edge
lying against the vertebræ, and the anterior thin margin
corresponding to the lower edge of the chest.   Observe
the three lobes of the liver;—the great or right lobe;[1]—
the small or left lobe;—the lobulus spigelii;—the great
fissure, separating the right and left lobe, and receiving
the suspensory ligament, and the ligamentum teres;—
the cavity of the portæ between the great lobe and lobulus spigelii;—the fissure on the right side of the lobulus
for the vena cava inferior, which fissure is almost a complete foramen;—the notch in the back part for the vertebræ;—the depression in the right lobe for the gall
bladder.   Observe the vessels in the cavity of the portæ,
the hepatic artery on the left side, the ductus communis
choledochus on the right side, and betwixt, but at the
same time behind them, the vena portæ; they are all
surrounded by a plexus of nerves. From the sympathetic and par vagum these vessels and nerves pass along

---

[1] Two others are enumerated, the lobulus caudatus and quadratus.
See Special Anatomy.

the edge of the little omentum, surrounded and connected by adipose and cellular substance; the part is called CAPSULA GLISSONI. Observe that the ligamentum teres was the umbilical vein of the fœtus, entering the vena portæ, and that the ductus venosus in the fœtus (obliterated in the adult), leaving the vena portæ, passed into one of the venæ cavæ hepaticæ.

3. VESICULA FELLIS, the GALL-BLADDER.—*Situation*. In the right hypochondrium, in a superficial depression on the under surface of the right lobe of the liver: it sends off the DUCTUS CYSTICUS, which, uniting with the DUCTUS HEPATICUS, forms the DUCTUS COMMUNIS CHOLEDOCHUS; this perforates the first curvature of the duodenum.

4. VENTRICULUS, the STOMACH.—*Situation*. In the left hypochondriac and epigastric regions. *Connected* to part of the interior surface of the diaphragm, to the concave surface of the liver by the little omentum, to the spleen by a reflection of peritoneum, and to the arch of the colon by the great omentum. Observe its greater curvature looking downward, its lesser curvature looking upward; and its two lateral surfaces. In the living body, the greater curvature is turned forward, and a little downward, the lesser arch backward, *i.e.* toward the spine, while one of the lateral convex sides is turned upward, and the other downward. Observe the bulging extremity on the left side, the cardia or upper orifice, where the œsophagus enters half-way between this great extremity and the lesser arch: the pylorus, or lower orifice, at the end of the small extremity, situated under the liver, and rather to the right side of the spine, feeling hard when touched. A constriction may be seen where the stomach ends in the duodenum, it marks the position of the pyloric valve in the inside.

5. The INTESTINES.—These form one continuous tube, but are divided into two portions, differing in their figure, structure, and functions, and distinguished by the names of small and large.

The small intestine is divided into duodenum, jejunum, and ileum; the large into cæcum, colon, and rectum.

Fig. 47.

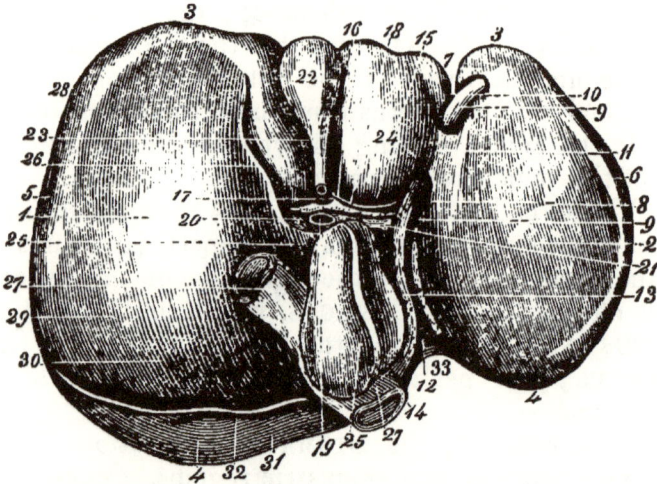

UNDER OR CONCAVE SURFACE OF THE LIVER.

1. Right Lobe.
2. Left Lobe.
3. Its Anterior or Inferior Edge.
4. Its Posterior or Diaphragmatic Portion.
5. Right Extremity.
6. Left Extremity.
7. Notch in the Anterior Margin.
8. Umbilical or Longitudinal Fissure.
9. Round Ligament or Remains of the Umbilical Vein.
10. Portion of the Suspensory Ligament in connection with the Round Ligament.
11. Pons Hepatis, or Band of Liver across the Umbilical Fissure.
12. Posterior End of Longitudinal Fissure.
13, 14. Attachment of the obliterated Ductus Venosus to the Ascending Vena Cava.
15. Transverse Fissure.
16. Section of the Hepatic Duct.
17. Hepatic Artery.
18 Its Branches.
19. Vena Portarum.
20. Its Sinus, or Division into Right and Left Branches.
21. Fibrous Remains of the Ductus Venosus.
22. Gall-bladder.
23. Its Neck.
24. Lobulus Quartus.
25. Lobulus Spigelii.
26. Lobulus Caudatus.
27. Inferior Vena Cava.

28. Curvature of Liver to fit the Ascending Colon.
29. Depression to fit the Right Kidney.
30. Upper Portion of its Right Concave Surface over the Renal Capsule.
31. Portion of the Liver uncovered by the Peritoneum.
32. Inferior Edge of the Coronary Ligament in the Liver.
33. Depression made by the Vertebral Column.

(1) SMALL INTESTINE;—about four times the length of the body.

a. The DUODENUM is broader than any other part of the small intestine, but is short; it takes a turn from the pylorus upward, and to the right side, passing under the liver and gall-bladder; then, turning upon itself, it descends, passing as low as the right kidney; it is in this space that it receives the pancreatic and gall ducts; thence it crosses before the renal vessels, before the aorta, and upon the last dorsal vertebra, firmly bound down by the peritoneum, which covers only its anterior surface; it then ascends from right to left, till it is lost under the root of the mesocolon.

Turning back the colon and omentum, fixing them over the brim of the thorax, and pushing down the small intestine toward the pelvis, you find the duodenum coming out from under the mesocolon, but still tied close to the spine; it terminates in the jejunum, exactly where the mesentery begins. The intestine in this course forms nearly a circle, the root of the mesocolon being the only part lying between its two extremities.

You have now to trace the rest of the small intestine, which lies convoluted in the umbilical and hypogastric regions.

b. JEJUNUM constitutes the first or upper half of the remaining small intestine, and is situated more in the upper part of the abdomen; it is redder, and its coats feel thicker to the touch, from the greater number of the valvulæ conniventes on its inner surface; its diameter exceeds that of the ileum.

c. The lower half is named ILEUM; it is situated more in the lower part of the abdomen, and terminates in the great intestine, by entering the caput coli, or beginning of the colon.

Fig. 48.

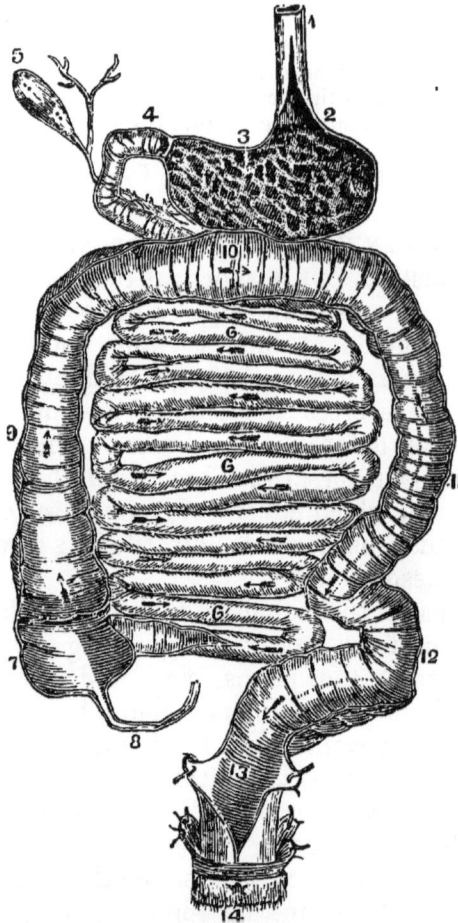

THE DIGESTIVE TUBE, FROM THE ŒSOPHAGUS TO THE ANUS.

1. Œsophagus, which is laid open at 2, to show its termination in the cardiac orifice of the stomach.
3. Interior of the Stomach with its rugæ.
4. Duodenum, commencing at the Pylorus.
5. Gall-bladder with the Cystic Duct, which last passes downward to open into the duodenum.
6, 6, 6. Small Intestine, terminating in 7, the Cæcum.
8. Appendicula Vermiformis.
9. Right ascending Colon.
10. Transverse Arch of the Colon.
11. Left descending Colon.   13. Rectum.
12. Sigmoid Flexure.   14. Anus.

As a general observation it may be said that the convolutions of the small intestine occupy the middle of the umbilical and hypogastric regions; but their situation varies much, particularly according to the state of the bladder and rectum. The course of the tube, independently of its convolutions, is from the left lumbar region, where the duodenum emerges from under the mesocolon to the right inguinal region, where the ileum terminates in the caput coli. Here a duplicature of the internal membrane into two folds, forms the ileo-cæcal valve.

(2) Great intestine.

d. The CÆCUM, or blind gut, is tied down by the peritoneum to the loins on the right side, lying in the space under the right kidney, hid by the convolutions of the ileum. On its posterior part there is a little appendage, of the shape of an earth-worm, named *Appendix Cæci Vermiformis.*

e. The COLON.—Its commencement at the Cæcum is termed CAPUT COLI, or head of the colon; it mounts upward from the cæcum over the anterior surface of the right kidney, passes under the gall-bladder, which, after death, tinges it with bile; and then, going across the upper part of the belly, forms its GREAT TRANSVERSE ARCH. In its whole course it is contracted into cells by its muscular fibres, which are united together, forming longitudinal bands; and it has some fatty projections attached to its surface, named Appendices Epiploicæ. Both these circumstances distinguish the large from the small intestine; which the difference of size does not always. The colon then goes backward under the stomach and spleen into the left hypochondrium; and then, descending over the left kidney, it is again tied down; it afterward turns over the brim of the pelvis, being at this part unconfined, and forming a loose and remarkable curvature, which is named the SIGMOID or ILIAC FLEXURE. After this convolution, the intestine assumes the name of

The RECTUM.—Drawing aside the intestines, you find the gut continued over the anterior surface of the sacrum and os coccygis to the anus.

Fig. 49.

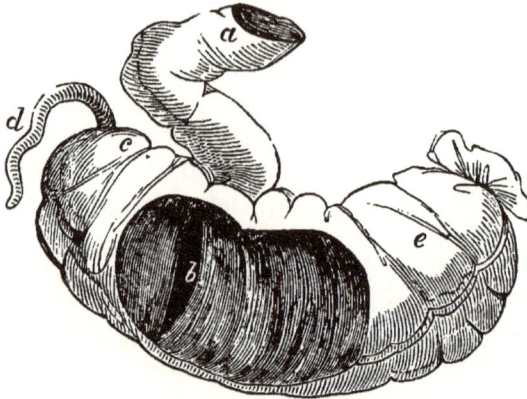

ILEO-CÆCAL VALVE.

a. The Terminal Part of the Ileum.
b. The Ileo-Cæcal Valve.
c. The Cæcum.
d. The Appendicula Vermiformis.
e. The Commencement of the Colon.

On pulling the stomach toward the right side, you will perceive

6. The LIEN, or SPLEEN.—*Situation.* In the left hypochondriac region, between the great extremity of the stomach, and the neighboring false ribs, under the edge of the diaphragm, to all of which it is connected by the peritoneum. It is of an oval figure; its external surface is gently convex; its internal surface irregularly concave, and divided by a longitudinal fissure, into which its vessels enter.

7. The PANCREAS.—*Situation.* This organ was in part seen on removing the little epiploon; it is more fully exposed by tearing through the great epiploon, between the great curvature of the stomach and the transverse arch of the colon. It lies in the cavity into which the foramen of Winslow leads; it extends from the fissure of the spleen across the spine, under the posterior surface of the stomach, and terminates within the circle formed by the duodenum; it is only covered on its anterior surface by the peritoneum.

The PANCREATIC DUCT pierces the coat of the duodenum, and enters the cavity of that intestine by an orifice common to it and to the ductus communis choledochus. The duct runs along the very centre of the gland, where the whiteness of its coats will readily enable the student to distinguish it.

All the abdominal viscera may now be removed, except the rectum, where it descends into the pelvis, which, being tied, should be allowed to remain, for it belongs to the demonstration of those parts; or the liver and its vessels, with the pancreas, may be left, and the vessels entering the portæ of the liver traced.

The student should open the different portions of the intestine, and mark the folds of the mucous membrane, VALVULÆ CONNIVENTES; also, the small eminences over those folds, VILLI. He should also examine the PYLORIC VALVE; the ILEO-CÆCAL VALVE, at the termination of the ileum. Little bodies are frequently seen under the mucous membrane of the duodenum, the GLANDS of BRUNNER, and other patches of ductless glands in the ileum, the GLANDS of PEYER.

The peritoneum should now be carefully dissected from the diaphragm, and from the sides and back part of the abdomen; thus the parts which lie more immediately behind that membrane may be examined.

8. RENES, the KIDNEYS.—Two glandular bodies, situated in the posterior part of the cavity of the abdomen, on each side of the lumbar vertebræ, between the last false rib and the spine of the ileum, and imbedded in a quantity of adipose membrane.

Observe the renal or emulgent artery entering the vein and ureter passing out of its fissure. Observe the course of the ureter; it passes behind the peritoneum over the psoas muscle into the pelvis, and runs between the rectum and bladder, which last it enters.

If the kidney be laid open from its convex to its concave margins, the following points may be observed: An exterior or CORTICAL PORTION; an internal, which being arranged in cones, formed of uriniferous tubes, the TUBULAR or CONICAL part. These cones look into

8*

three cavities, INFUNDIBULA, and these again into the PELVIS, and this into the ureter.

Fig. 50.

LONGITUDINAL SECTION OF THE KIDNEY, WITH ITS RENAL CAPSULE.

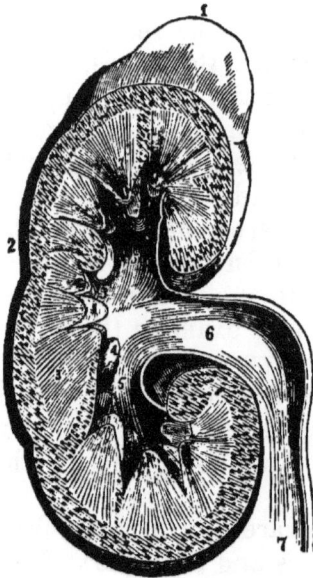

1. Renal Capsule.
2. Cortical or vascular part of the Kidney.
3, 3. Uriniferous Tubes collected into conical Fasciculi.
4, 4. Papillæ, projecting into their corresponding calices.
5, 5, 5. The Three Infundibula.
6. Pelvis of the Kidney.
7. Ureter.

9. The CAPSULÆ RENALES.—Two glandular bodies, situated on the upper extremity of each kidney, of an irregular figure, crescent-like, or somewhat triangular.

By the removal of the peritoneum, several muscles are exposed, situated at the superior and posterior parts of the abdomen.

One single muscle is situated in the superior part of the abdomen.

DIAPHRAGMA, the DIAPHRAGM, or MIDRIFF.—This is a broad, thin, muscular septum between the thorax and abdomen, situated obliquely; it is concave below, and convex above, the middle of it on each side reaching as high within the thorax as the fourth rib. It is divided into two portions.

1. The superior or greater muscle of the diaphragm

forms the transverse partition between the chest and abdomen,

Arising, by distinct fleshy fibres, 1. From the posterior surface of the ensiform cartilage; 2. From the cartilages of the seventh, and all the false ribs; 3. From the ligamentum arcuatum, which is a ligament extended, somewhat indistinctly, from the top of the twelfth rib to the lumbar vertebræ, forming an arch over the psoas and quadratus lumborum muscles. From these origins the fibres run, in different directions, like radii from the circumference to the centre of a circle, and are

Inserted into a broad tendon (*tendinous centre*, or *cordiform tendon*), which is situated in the middle of the diaphragm, and in which, therefore, the fibres from the opposite sides are interlaced.

2. The inferior or lesser muscle, or appendix of the diaphragm, lies on the bodies of the vertebræ, and

Arises, by four small tendinous feet, on each side, from the second, third, and fourth lumbar vertebræ; these tendons soon join, to form a strong pillar on each side, named the Crus of the Diaphragm. The crura run obliquely upward and forward, form two fleshy bellies, a fasciculus of each of which crossing over to the other, decussates with the opposite one, and thus forms the interval of the two crura into a superior and inferior opening.

Inserted into the posterior part of the middle cordiform tendon.

Situation. The diaphragm is covered on its superior surface by the pleura, and on its inferior by the peritoneum; it separates the thoracic from the abdominal viscera. It is perforated in its fleshy and tendinous parts by several bloodvessels, and other important organs.

(1) The aorta passes between the tendinous part of the crura, lying close upon the spine; and the thoracic duct passes betwixt the aorta and the right crus.

(2) A little above, and to the left side of the aorta, the œsophagus, with the eighth pair of nerves attached to it, passes through an oval fissure formed in the fleshy columns of the inferior muscle.

(3) The vena cava perforates the tendon toward the right side.

Fig. 51.

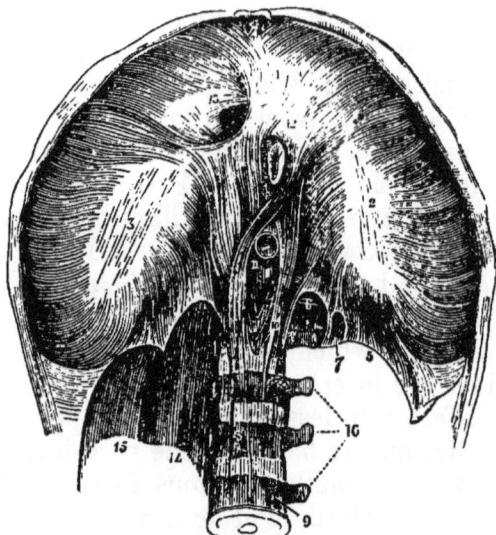

THE DIAPHRAGM.

1, 2, 3. Tendinous Centre of the Greater Diaphragm.
5, 6. Ligamentum Arcuatum.
7. Foramen of the Lesser Splanchnic Nerve.
8. Right Crura of Diaphragm.
9. Fourth Lumbar Vertebra.
10. Left Crura of Diaphragm.
11. Hiatus Aorticus.
12. Foramen Œsophageum.
13. Foramen Quadratum, for the Passage of the Vena Cava.
14. Psoas Muscle.
15. Quadratus Lumborum.
16. Transverse Processes of the Lumbar Vertebræ.

(4) The great splanchnic nerve, and branches of the vena azygos vein, perforate some of the posterior fibres of the crura. The lesser splanchnic nerve also passes through an opening in the substance of the crura.

(5) On each side of the sternum there is a small fissure, where the peritoneum and pleura are only separated by adipose membrane.

Use. The diaphragm is one of the chief agents in respiration; it also acts in coughing, laughing, and speaking, and in the expulsion of the urine and feces, etc.

Four pair of muscles are situated within the posterior part of the cavity of the abdomen.

1. The PSOAS PARVUS, often wanting.—It *arises*, fleshy, from the sides of the last dorsal and first lumbar vertebræ; it sends off a small long tendon, which, running on the inside of the psoas magnus, is

Inserted, thin and flat, into the brim of the pelvis, at the junction of the os ilium and pubis.

Use. To assist the psoas magnus in bending the loins forward.

2. The PSOAS MAGNUS.—It *arises*, fleshy, from the

Fig. 52.

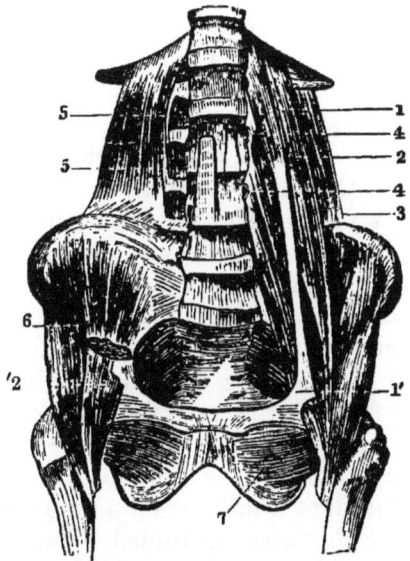

1. Small Psoas Muscle.
1'. Insertion of the Tendon of the same into the Iliac Fascia cut.
2. Great Psoas Muscle.
3. Quadratus Lumborum Muscle, partly concealed by the two Psoas Muscles.
3'. Same of the Right Side entirely exposed
4, 4. Foramina formed by the Grooves upon the Bodies of the Lumbar Vertebræ, and the Origins of the Great Psoas Muscle, for the passage of the Lumbar Arteries and Veins.
5, 5. Inter-Transverse Muscles.
6. Iliac Muscle entirely exposed by the removal of '2, Great Psoas Muscle cut.

side of the body, and transverse process of the last vertebra of the back, and in the same manner from all those of the loins, by as many distinct slips. It runs down over the brim of the pelvis, and is

Inserted, tendinous, into the trochanter minor of the os femoris, and fleshy, into that bone immediately below the trochanter.

Situation. It is situated betwixt the psoas parvus and iliacus internus.

Use. To bend the thigh forward, roll it outward; or, to assist in bending the body.

3. The ILIACUS INTERNUS.—It *arises*, fleshy, from the transverse process of the last vertebra of the loins, from all the inner margin of the spine of the os ilium, from the edge of that bone between its anterior superior spinous process and the acetabulum, and from all its hollow part between the spine and the linea innominata. Its fibres descend under the outer half of Poupart's ligament, and join the tendon of the psoas magnus.

Inserted with the psoas magnus.

Situation. It fills up the internal concave surface of the os ilium. It is covered by a pretty strong fascia, which is inserted into the crista of the ilium, and into the crural arch, *Fascia Iliaca:* at which point it joins the Fascia Transversalis. This ILIAC FASCIA passes under the iliac bloodvessels into the pelvis. The latter insertion prevents the bowels from descending under Poupart's ligament, except at the inner edge of the iliac vein, which is accordingly the situation of a crural hernia, and which the student should examine most carefully. A part of this fascia is also continued behind the femoral vessels over the pubis, to form a part of the sheath which incloses those vessels. A short distance to the pubic side of the iliac vein, a strong semilunar tendinous edge is seen; this is one of the attachments of Poupart's ligament to the linea innominata, and is called *Gimbernat's ligament.* Between this and the femoral vein is a space filled by a lymphatic gland and cellular tissue. The last called the *Septum Crurale.* This space is the *Femoral* or *Crural ring*, and is the place where an intestine some-

times passes through, forming Femoral Hernia. The septum crurale would form one of its coverings. Remember
these points, and apply them when conducting the dissection of the thigh.

Use. To assist in bending the thigh, and in bringing
it directly forward.

N. B. The insertion of the two last described muscles
cannot be seen till the thigh is dissected.

4. The QUADRATUS LUMBORUM.—This muscle *arises*,
tendinous and fleshy, from rather more than the posterior
third part of the spine of the os ilium.

Inserted into the transverse processes of all the vertebræ of the loins, into the posterior half of the last rib,
and, by a small tendon, into the side of the last vertebra
of the back.

Situation. It is situated laterally at the lower part of
the spine, more outwardly than the psoas magnus.

Use. To move the loins to one side, pull down the last
rib. If both act, to bend the loins forward.

Of the Vessels and Nerves situated behind the Peritoneum.

1. *The Arteries, viz.: The Aorta Abdominalis, and its branches.*

The Aorta passes from the thorax into the abdomen,
between the crura of the diaphragm, close upon the spine.
It then descends on the forepart of the vertebræ, inclined
to the left side. On the fourth lumbar vertebra, it bifurcates into the two primitive or common iliac arteries.

BRANCHES OF THE ABDOMINAL AORTA.—1. The two
PHRENIC Arteries arise from the Aorta, before it has
fairly entered into the abdomen, and ramify over the
diaphragm.

2. The CŒLIAC Artery or Axis comes off at the point
where the aorta has fairly extricated itself from the diaphragm, surrounded by the meshes of the semilunar ganglion. It divides into three branches.

(1) A. GASTRICA, smallest of the three. It passes along

the lesser curvature from left to right, to inosculate with
the pylorica or coronaria dextra.

Fig. 53.

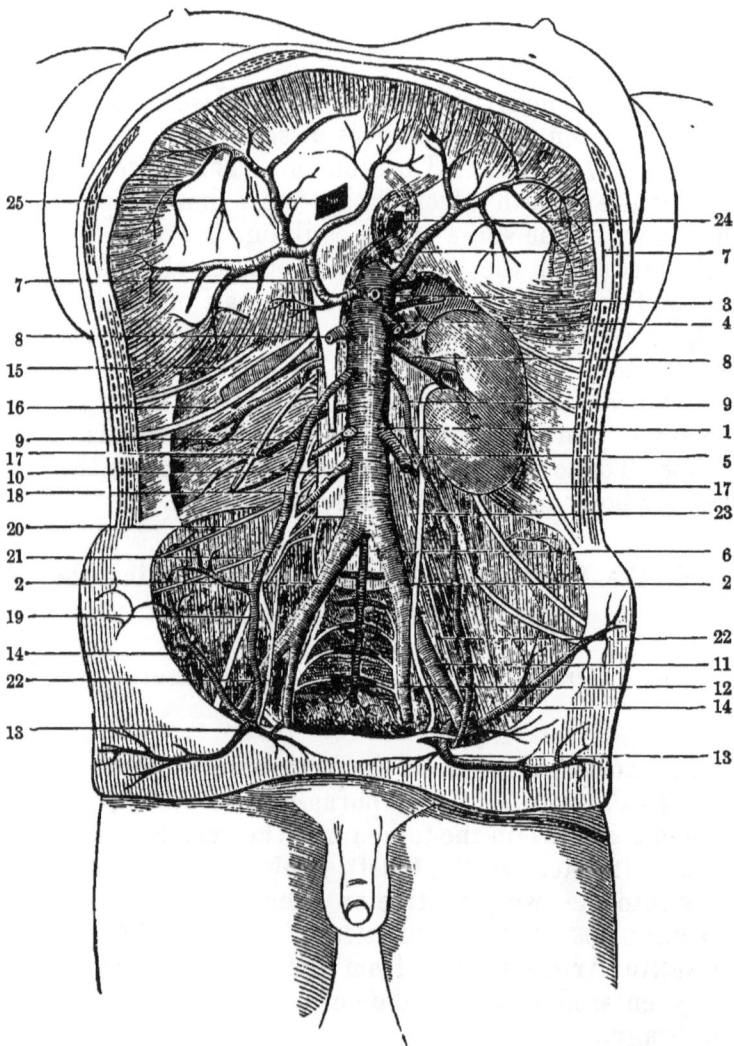

ABDOMINAL PORTION OF THE AORTA AND ITS BRANCHES.

1. Aorta.
2, 2. Primitive Iliac Arteries.
3. Cœliac Artery cut across.

4. Superior Mesenteric cut.
5. Inferior Mesenteric cut.
6. Middle Sacral.

7, 7. Diaphragmatic or Phrenic
    Arteries.
8, 8. Renal Arteries.
9, 9. Spermatic Arteries.
  10. Lumbar Arteries.
  11. External Iliac Artery.

12. Internal Iliac Artery.
13, 13. Epigastric Artery.
14. Circumflex Iliac Artery.
15, 16. Musculo-cutaneous
    Nerves.

(2) ARTERIA SPLENICA passes under the stomach, and along the upper border of the pancreas; it enters the spleen, and gives off the following branches:

a. PANCREATICÆ PARVÆ to the pancreas, where it runs along the border of that viscus.

b. VASA BREVIA to the bulging extremity of the stomach.

c. A. GASTRO-EPIPLOICA SINISTRA, along the greater curvature of the stomach, inosculating with the gastro-epiploica dextra.

(3) ARTERIA HEPATICA runs to the liver. It sends off the following branches:

a. PYLORICA.—It sends its ramifications along the lesser curvature, to inosculate with the proper coronary artery.

b. GASTRO-EPIPLOICA DEXTRA, or GASTRO-DUODE-NALIS, passes under the pylorus, and along the great curvature of the stomach, inosculating with the gastro-epiploica sinistra (from the splenic), and sends off a branch to the pancreas.

The hepatic artery then divides into the right and left hepatic. The RIGHT is distributed to the right lobe of the liver, and to the gall-bladder. The LEFT supplies the whole of the left lobe, the lobulus spigelii, and part of the right lobe of the liver.

3. The SUPERIOR MESENTERIC ARTERY.—It leaves the aorta about half an inch lower than the cœliac artery; it enters the fold of peritoneum forming the mesentery, and runs down in this, incurvating from the left to the right side.

From the right side or concavity of this arch, three branches are given to the colon.

(1) A. ILEO-COLICA to the caput coli and last of the ileum.

(2) A. COLICA DEXTRA to the right side of the colon.

(3) A. COLICA MEDIA to the arch of the colon.

· Fig. 54.

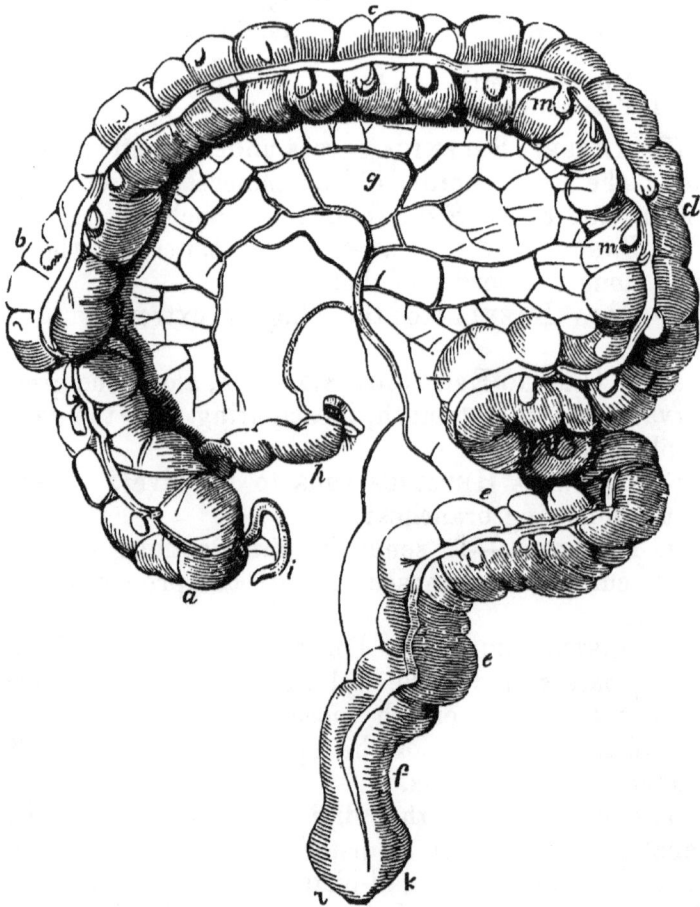

THE SEVERAL PARTS OF THE LARGE INTESTINE.

   a. The Cæcum.
   b. Right or Ascending Colon.
   c. Transverse Colon, or Arch of the Colon.
   d. Left or Descending Colon.
e, e. Sigmoid Flexure of the Colon.
   f. Rectum.
   g. Mesocolon.
   h. The end of the Ileum, or its termination at the Ileo-cæcal Valve.
   i. Appendicula Vermiformis.
   k. Pouch of the Rectum.
   l. Anus.
   m. Appendices Epiploicæ.

The convexity of the arch of the superior mesenteric sends off from sixteen to twenty branches, which, forming frequent anastomoses and arches, proceed to the small intestines.

4. The RENAL or EMULGENT ARTERIES are two in number. Below the superior mesenteric, pass to the kidney. The right artery longer than the left, and passes behind the vena cava ascendens.

5. The SPERMATIC ARTERIES are also two; they come off an inch below the emulgent from the forepart of the aorta. Each artery descends in its course, accompanied by the spermatic vein and nerves. It passes through the abdominal rings, and enters the upper part of the testicle. In the female it supplies the ovaria and fundus uteri.

6. The INFERIOR MESENTERIC is a single trunk. Comes off rather from the left side of the aorta, below the spermatic arteries; it passes in the mesentery to the left side of the abdomen, where it divides as follows:

(1) The COLICA SINISTRA. To the left side of the colon. Inosculates with the A. Colica Media.

(2) Branches which pass to the sigmoid flexure.

(3) The great trunk of the artery runs down to the rectum, on which it ramifies. Is termed ART. HEMORRHOIDALIS SUPERIUS.

7. The LUMBAR ARTERIES are five or six small arteries on each side, which arise from the back part of the aorta, and are distributed to the spinal canal, muscles of the spine, and of the sides of the abdomen and pelvis.

8. A. SACRA MEDIA is a single artery, arises from the back part of the aorta at its bifurcation, and descends along the anterior surface of the sacrum, giving twigs to all the neighboring parts.

At the fourth lumbar vertebra, the aorta bifurcates into the two primitive or common iliacs.

The ILIACA COMMUNIS runs along the edge of the psoas muscle, and at an inch or two from its origin divides into

(1) The Internal Iliac which passes down into the pelvis.

(2) The External Iliac, which, following the direction of the psoas muscle, passes under Poupart's ligament, and becomes the Femoral artery.

2. Veins.

The VENA CAVA ABDOMINALIS, vel Inferior, is formed by the junction of the two common iliac veins; it passes up through the abdomen on the right side of the aorta.

In this course it receives the following veins, which resemble their corresponding arteries:

1. The Lumbar Veins.
2. The Emulgent or Renal Veins: the left is the longest, as it crosses over the forepart of the aorta.
3. The Right Spermatic Vein; the left enters the left renal vein.

The vena cava then passes through the fissure of the liver, being nearly surrounded by that viscus, and receiving three branches from it, called the VENÆ HEPATICÆ. It then perforates the diaphragm, and enters the thorax.

The common iliac vein of each side is formed by the union of two branches, the EXTERNAL and INTERNAL ILIAC VEINS, which accompany the arteries of the same name. The common iliac vein of each side lies on the inside of its artery; hence both veins cross behind the right iliac artery, to unite and form the vena cava, on the forepart of the lumbar vertebræ.

The SUPERIOR MESENTERIC VEIN, the INFERIOR MESENTERIC VEIN, and the SPLENIC VEIN do not join the cava, but are united behind the pancreas, to form the Vena Portæ. This vein ramifies anew through the liver, and its blood is returned into the vena cava by the venæ cavæ hepaticæ.

Although the trunks just enumerated are the chief veins that contribute to form the vena portæ, yet the returning veins of all the viscera inclosed in the peritoneum, except the liver, are included in the same system, and join one or other of the large trunks. This is the case with the stomach, pancreas, gall-bladder, and omentum. The blood which goes to the spleen, large and small intestines, is all returned by the three great trunks.

3. Nerves.

1. The eighth pair, or Par Vagum, descending on each side of the œsophagus through the diaphragm, forms the two STOMACHIC PLEXUSES on the anterior and posterior surfaces of the stomach. These plexuses send some branches to the cœliac, to the hepatic, and to the splenic plexus.

2. The SPLANCHNIC NERVE, or Anterior Intercostal, a branch sent off by the intercostal nerve in the thorax, enters the abdomen betwixt the crura of the diaphragm; here each nerve forms a SEMILUNAR GANGLION by the side of the cœliac artery.

From the ganglion on each side branches are sent across, which communicate intimately together, and form round the root of the cœliac artery a very intimate plexus, containing several ganglia of various sizes, formerly called the SOLAR or CŒLIAC PLEXUS. Nerves pass from this plexus, with the branches of the aorta, to the various viscera of the abdomen; they form the HEPATIC, SPLENIC, SUPERIOR and INFERIOR MESENTERIC, RENAL, and SPERMATIC PLEXUSES.

3. The trunk of the sympathetic nerve (the *posterior*) perforates the diaphragm close to the spine, and runs along the upper edge of the psoas magnus. It terminates on the extremity of the os coccygis by union with the nerve of the opposite side, in a ganglion (GANGLION IMPAR). In this course it communicates with the lumbar nerves and the various abdominal plexuses.

The THORACIC DUCT may be seen passing from the abdomen into the thorax, between the aorta and the right crus of the diaphragm. It is larger here than in its subsequent course. It empties into the junction of the left subclavian and internal jugular veins.

CHAPTER XI.

DISSECTION OF THE ANTERIOR PART OF THIGH.

CARRY an incision from the middle of Poupart's ligament obliquely across the thigh, and around the internal condyle of the femur to the tubercle of the tibia. Reflect the integuments. The superficial fascia, especially at the groin, is lamellated; and situated in it about Poupart's ligament a cluster of lymphatic glands. In and beneath this fascia notice

1. The VENA SAPHENA MAJOR, seen running up in the inside of the knee and thigh. At first it lies very superficial, betwixt the skin and fascia lata. As it ascends it is gradually enveloped by the fibres of the fascia, and then sinks beneath it to join the femoral vein about an inch below Poupart's ligament. The space through which it sinks to join the femoral vein is the SAPHENOUS OPENING, and the loose cellular tissue which envelops the vein and occupies the opening is called the CRIBRIFORM FASCIA. In its course it is joined by several cutaneous veins.

2. Immediately under the true skin you may occasionally perceive the LYMPHATIC VESSELS running, like lines of a whitish color, to enter the inguinal glands.

3. Several CUTANEOUS NERVES are seen ramifying above the fascia. They all come from the lumbar nerves or anterior crural nerve.

The deep fascia, or FASCIA LATA, may now be exposed distinctly by carefully clearing away the superficial fascia; preserving, however, the saphena vein, a short distance down. Observe how extensively it arises from the bones, tendons, and ligaments. On the anterior and superior part of the thigh, it arises from Poupart's ligament, from the os pubis, from the descending ramus of that bone, and from the ascending ramus and tuberosity of the ischium; behind, and on the outside, from the

Fig. 55.

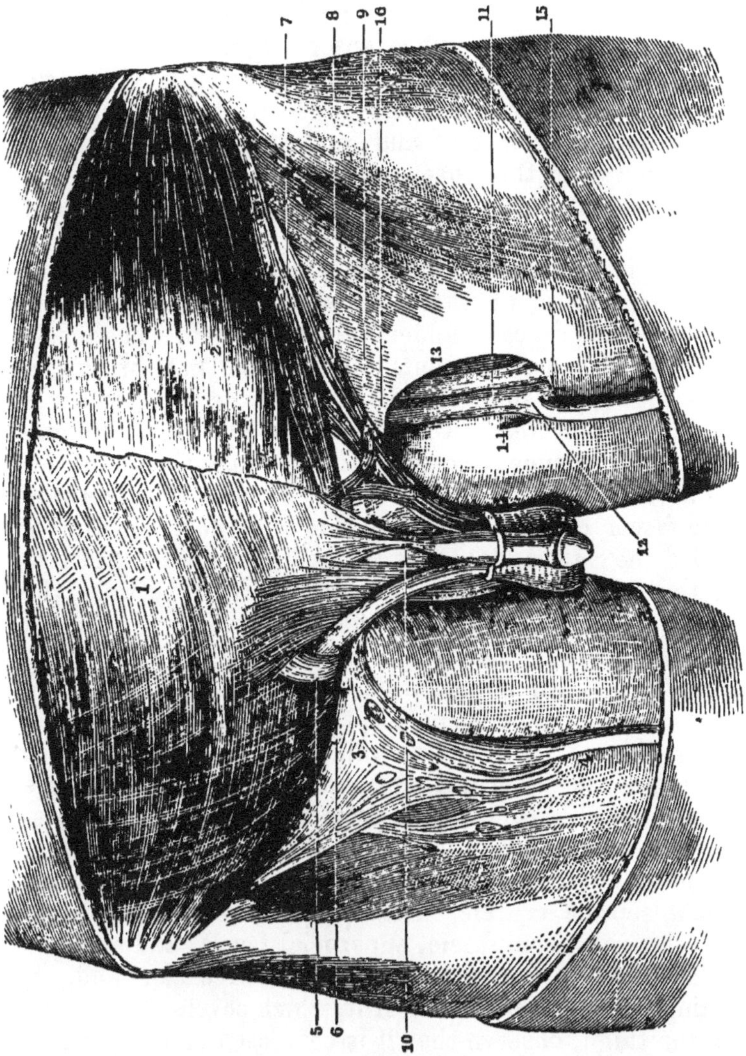

DISSECTION OF SOME OF THE PARTS CONCERNED IN FEMORAL AND
INGUINAL HERNIA.

1. Tendon of the External Oblique Muscle.
2. Tendon of the Internal Oblique, the first-named muscle being
   dissected off.
3. Cribriform Fascia.
4. Vena Saphena.
5. External Abdominal Ring and Spermatic Cord.
6. Poupart's Ligament.

whole spine of the ilium, and from the sacro-sciatic ligaments. It receives a number of fibres from a muscle
belonging to it, viz., the tensor vaginæ femoris, and
from the tendon of the gluteus-maximus; it passes down
over the whole thigh, is firmly fixed to the linea aspera,
to the condyles of the femur, and to the patella, and is
continued over the knee, to be attached to the heads of
the tibia and fibula, after which it forms the fascia of
the leg.

On the upper and anterior part of the thigh, below
Poupart's ligament, there is a slight hollow, where the
great vessels descend under the crural arch. The fascia
lata forms, just on the outside and upper part of this, a
crescent-shaped fold, called its *semilunar edge*, which is
strongly connected to the crural arch and linea innominata. The commencement of this semilunar edge, sometimes called the Superior Cornu, is HEY'S LIGAMENT.
The fascia on the outside of the saphenous opening is
called the SARTORIAL, that on the inside the PECTINEAL.
If the cribriform fascia be removed from between the
two, we will see the femoral bloodvessels lying beneath
inclosed in their sheath. Press the finger down from
the inside of the abdomen on the pubic side of the iliac
vein, and it will be found to project in the thigh on the
inner side of the femoral vein, but within its sheath.
This space is the crural ring, and gives you the course
of a femoral hernia, which projects forward, after passing under Poupart's ligament through the saphenous
opening, covered, as you may perceive, by the peritoneum, septum crurale, fascia propria, or sheath of the
vessels, cribriform fascia, superficial fascia, and skin.

The fascia should now be dissected back; and, in
lifting up the thicker part of it, which covers the outside
of the thigh, observe that it is composed of two laminæ
of fibres. The fibres of the outer lamina run in circles
round the thigh, while those on the inside, which are
stronger and more firmly connected, run longitudinally.

Fig. 56.

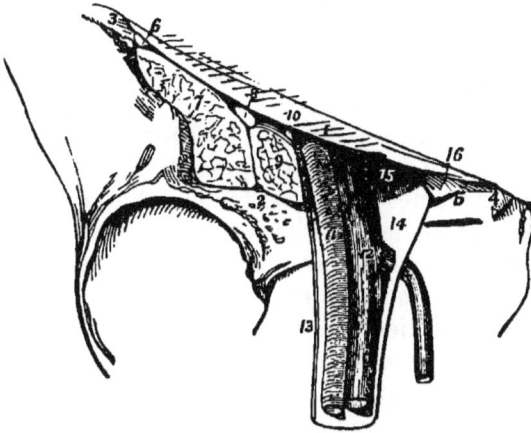

THE FEMORAL OR CRURAL ARCH, AND THE STRUCTURES SITUATED BE-
TWEEN IT AND THE ANTERIOR PART OF THE SUPERIOR MARGIN OF
THE PELVIS.

1. The Crural Arch, or Poupart's Ligament.
2. Pubic Bone.
3. Superior Anterior Spine of the Ilium.
4. Spine of the Pubis.
5. Pectineal Line, and the Insertion of Gimbernat's Ligament.
7. Iliac Muscle cut.
8. Crural Nerve cut.
9. Great Psoas Muscle cut.
10. Point at which the Crural Branch of the Genito-Crural Nerve
    reaches the Thigh.
11. Femoral Artery.
12. Femoral Vein, receiving the Saphena Vein.
13. External Portion of the Sheath of the Femoral Vessels lying in
    contact with the Femoral Artery.
14. The large funnel-shaped Cavity of the Sheath on the Inner Side
    of the Femoral Vein.
15. Internal Femoral Ring, bounded *above* by the Crural Arch, *be-
    hind* by the Pubis, *externally* by the Vein, and *internally* by
    the free edge of (16) Gimbernat's Ligament.

Muscles Situated on the Forepart and Inside of the Thigh.

These are nine in number.

1. The TENSOR VAGINÆ FEMORIS—*Arises*, by a narrow, tendinous, and fleshy origin, from the external part
of the anterior superior spinous process of the os ilium;
it forms a considerable fleshy belly.

Inserted into the inner side of the great fascia, where it covers the outside of the thigh, and a little below the trochanter major.

Use. To stretch the great fascia of the thigh, and assist in its abduction.

2. The SARTORIUS—*Arises,* by short tendinous fibres, from the anterior superior spinous process of the os ilium, soon becomes fleshy, extends obliquely across the thigh, and passes behind the inner condyle.

Inserted, by a broad and thin tendon, into the inner side of the tibia, immediately below its anterior tubercle.

Situation. It lies before the muscles of the thigh, crossing them like a strap about two inches in breadth. At the lower part of the thigh, it runs between the tendon of the triceps adductor magnus, and that of the gracilis. It is inserted above tendons of the gracilis and semitendinosus, over which it sends an aponeurotic expansion.

Use. To bend the leg obliquely inward on the thigh, and to bend the thigh forward.

It crosses over the femoral artery.

3. The RECTUS FEMORIS—*Arises,* by a strong tendon, from the inferior anterior spinous process of the os ilium; and, by another strong tendon, from the dorsum of that bone a little above the acetabulum, and from the capsular ligament of the hip-joint. The two tendons soon unite, and send off a large belly, which runs down over the anterior part of the thigh, and terminates in a flat strong tendon, which is

Inserted into the upper extremity of the patella.

Situation. To expose the tendinous origins of this muscle, the origins of the sartorius and tensor vaginæ femoris must be raised. Its insertion lies betwixt the two vasti.

Use. To extend the leg on the thigh, and to bend the thigh on the pelvis; to bring the pelvis and thigh forward to the leg.

Under the rectus, and partly covered by it, there is a large mass of flesh, which, at first sight, appears to form but one muscle. It may, however, be divided into three;

the separation on the external surface is not generally very evident, but, by following the course of the vessels which enter this mass, and by cutting through perhaps a few fibres externally, you will discover the line of separation; and this separation, as you proceed deeper with your dissection, will become very distinct. The three

Fig. 57.

A VIEW OF THE MUSCLES ON THE FRONT OF THE THIGH.

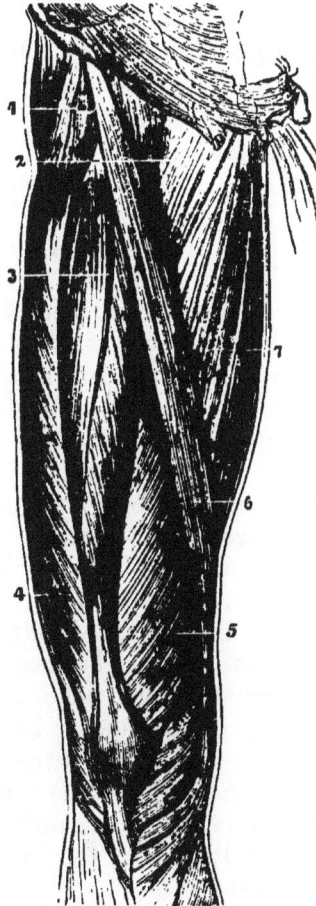

1. Tensor Vaginæ Femoris.
2. Pectineus.
3. Rectus Femoris.
4. Vastus Externus.
5. Vastus Internus.
6. Sartorius.
7. Adductor Longus.

muscles are named vastus externus, vastus internus, and cruræus: at the upper and middle parts of the thigh, they may be separated very distinctly; but for two or three inches above the condyles they are connected inseparably.

4. The VASTUS EXTERNUS—*Arises*, tendinous and fleshy, from the anterior surface of the root of the trochanter major, from the outer edge of the linea aspera, its whole length, from the oblique line running to the external condyle, and from the whole external flat surface of the thigh-bone. The fleshy fibres run obliquely forward.

Inserted into the external surface of the tendon of the rectus cruris, and into the side of the patella.

Situation. This muscle forms the large mass of flesh on the outside of the thigh.

Use. To extend the leg, or to bring the thigh forward upon the leg.

5. The VASTUS INTERNUS—*Arises*, tendinous and fleshy, from the forepart of the root of the trochanter minor, from all the upper edge of the linea aspera, from the oblique line running to the inner condyle, and from the whole internal surface of the thigh-bone. Its fibres descend obliquely downward and forward.

Inserted into the lateral surface of the tendon of the rectus cruris, and into the side of the patella.

Situation. This muscle embraces the inside of the femur in the same manner as the last described muscle does the outside, but it is much smaller;—it is also in part covered by the rectus. At its upper part the sartorius passes over it obliquely.

Use. Same as the last.

6. The CRURÆUS, or CRURALIS—*Arises*, fleshy, from between the two trochanters of the os femoris, from all the forepart of the bone, and from the outside as far back as the linea aspera; but from the inside of the bone it does not arise, for between the forepart of the femur and the inner edge of the linea aspera, there is a smooth plain surface, of the breadth of an inch, extending nearly the whole length of the bone, from which no muscular fibres arise.

Inserted into the posterior surface of the tendon of the rectus, and upper edge of the patella.

Situation. The principal part of this muscle is lapped over, and concealed, by the bellies of the two vasti.

Use. Same as the last.

7. The GRACILIS—*Arises*, by a broad thin tendon, from the lower half of that part of the os pubis which forms the symphysis, and from the inner edge of the descending ramus: it soon grows fleshy, and forms a belly, which, becoming narrower as it descends, terminates in a tendon, which passes behind the inner condyle of the thigh-bone, and is reflected forward, to be

Inserted in the inside of the tibia.

Situation. From the pubis to the knee it runs immediately under the integuments on the inside of the thigh; it is inserted below the tendon of the sartorius, and above that of the semitendinosus.

Use. To bring the thigh inward and forward, and to assist in bending the leg.

8. The PECTINALIS—*Arises*, fleshy, from that ridge of the os pubis which forms the brim of the pelvis, and from the concave surface below the ridge: it forms a thick flat belly.

Inserted, by a flat tendon into the linea aspera, immediately below the lesser trochanter.

Situation. Its origin lies on the inside of the belly of the psoas magnus, where that muscle slides over the brim of the pelvis, and on the outside of the origin of the adductor longus.

Use. To bend the thigh forward, to move it inward, and to perform rotation, by turning the toes outward.

9. The TRICEPS ADDUCTOR FEMORIS consists of three distinct muscles, which, passing from the pelvis to the thigh, lie in different layers upon one another, and have nearly the same action.

(1) The ADDUCTOR LONGUS—*Arises*, by a short strong tendon, from the upper and inner part of the os pubis, near its symphysis;—forms a large triangular belly, which, as it descends, becomes broader, but less thick.

Inserted, tendinous, into the middle part of the linea aspera: from its tendon and that of the vastus internus, a membranous canal is formed for the femoral artery.

The ANASTOMOTICA MAGNA ARTERY is given off here also.

Fig. 58.

ANOTHER VIEW OF THE ADDUCTOR MUSCLES WITH THE PECTINEUS.

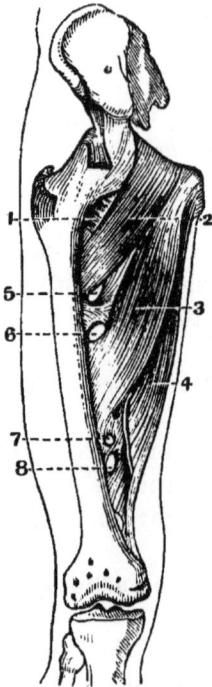

1. Upper Part of Adductor Magnus.
2. Pectineus.
3. Adductor Longus.
4. Adductor Magnus.
5, 6. Foramina for the First and Second Perforating Arteries.
7, 8. Foramina for the Femoro-popliteal Vessels.

Situation. It arises betwixt the pectinalis and gracilis, and above the adductor brevis.

(2) The ADDUCTOR BREVIS—*Arises*, fleshy and tendinous, from the os pubis, between the lower part of the symphysis pubis, and the foramen thyroideum ;—it forms a fleshy belly.

Inserted, tendinous, into the upper third of the linea aspera.

Situation. Its origin lies under the origins of the pectinalis and adductor longus, and on the outside of the tendon of the gracilis. It is inserted behind those muscles, but before the adductor magnus. The obturator nerve lies upon it after coming through the obturator foramen, with an artery of the same name.

Fig. 59.

ADDUCTOR MUSCLES, WITH THE OBTURATOR EXTERNUS.

1 Femur.
2. Ilium.
3. Pubis.
4. Obturator Externus.
5. Superior Fasciculus of the Adductor Magnus.
6, 7. Adductor Brevis
8. Adductor Longus.
9, 10. Adductor Magnus.
11. Foramen for the Passage of the Perforating Arteries.
12. Same for Femoro-popliteal Vessels.

(3) The ADDUCTOR MAGNUS — *Arises*, principally fleshy, from the lower part of the body, and from the descending ramus of the os pubis, and from the ascending ramus of the ischium, as far as the tuberosity of that bone. The fibres run outward and downward, having various degrees of obliquity.

Inserted, fleshy, into the whole length of the linea aspera, into the oblique ridge above the internal condyle of the os femoris, and, by a roundish long tendon, into the upper part of that condyle.

Situation. This large muscle arises behind and below the two other adductors; it forms a flat partition betwixt the muscles on the fore and back parts of the thigh.

Use. To approximate the thighs to each other;—to roll them outward.

Arteries, Veins, and Nerves on the Forepart and Inside of the Thigh.

1. *Arteries.*

The FEMORAL ARTERY may be said to pass along the inside of the thigh, where it emerges from under Poupart's ligament; it lies cushioned on the fibres of the psoas magnus, is called the inguinal artery, and is very nearly in the mid space between the angle of the pubis and the anterior superior spine of the ilium, nearer however by a finger's breadth to the former; having left the groin, it assumes the name of Femoral, and in its course down the thigh, runs over the following muscles: the pectinalis, part of the adductor brevis; the whole of the adductor longus, and about an inch of the adductor magnus; it then slips betwixt the tendon of the adductor magnus and the bone, and, entering the ham, becomes the Popliteal artery. There is a strong interlacing of the tendinous fibres, forming a deep groove for the artery between the adductor longus and vastus internus. The artery is accompanied by the Femoral vein, first inside, and then gradually getting behind it. The long saphenous nerves and short saphenous nerves accompany the vessels, as will presently be seen. It is also invested by a firm sheath, which consists of condensed cellular membrane, intermixed with some tendinous fibres : for some inches below Poupart's ligament, this artery is on its forepart only covered by cellular substance, absorbent glands, and the general fascia of the thigh; but, meeting with the inclined line of the sartorius, it is, during the rest of its course, covered by that muscle. It perforates the tendon of the adductor magnus, at the distance of rather more than one-third of the length of the bone from its lower extremity.

BRANCHES OF THE FEMORAL ARTERY.—The A. PRO-FUNDA comes off from the femoral artery at the distance

of two or three inches from Poupart's ligament ; it is
nearly as large as the femoral itself, runs down for some
little way behind it, and terminates in three or four

Fig. 60.

ARTERIES SEEN ON THE FRONT OF THE THIGH.

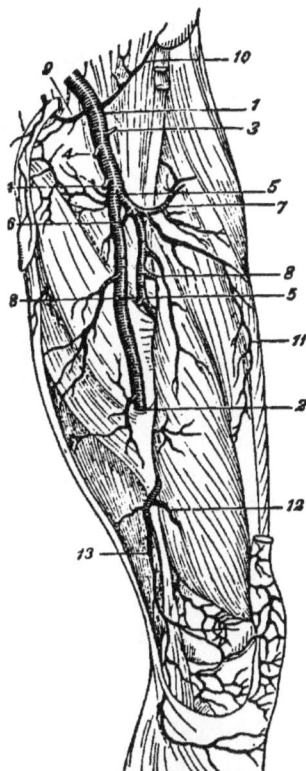

1, 2. Femoral Artery.
   3. Superficial Epigastric, cut off.
4, 4. External Pudics, cut off.
5, 5. Profunda Femoris.
   6. Internal Circumflex.
   7. External Circumflex.
8, 8. Perforating Arteries.
   9. Epigastric.
 10. Circumflexa Ilii.
 11. Muscular Branch.
 12. Superior Internal Articular Ar-
    tery.
 13. One of its Branches. The Pop-
    liteal Artery begins where the
    Femoral terminates, at 2.

branches, which, perforating the triceps adductor, are
named ARTERIÆ PERFORANTES. These supply the great
mass of muscles on the back part of the thigh, and inosculate largely with the sciatic, gluteal, and obturator
arteries. The profunda also sends off two considerable
branches (the *first ones*), which, encircling the upper
part of the thigh, are named CIRCUMFLEXA INTERNA,
and CIRCUMFLEXA EXTERNA.

The small branches of the femoral artery before the
profunda may be enumerated as follows: 1. Some twigs

9*

to the inguinal glands (*inguinal*); 2. Some to the external parts of generation, named Pudicæ Externæ; 3. One or two going toward the anterior superior spinous process of the ilium (SUPERFICIAL CIRCUMFLEX ILII); 4. One upward over the fascia of the abdomen, SUPERFICIAL EPIGASTRIC, or arteria ad cutem abdominis.

The OBTURATOR ARTERY, arteria obturatrix, which is a branch of the internal iliac artery, passes through the notch at the upper part of the foramen thyroideum, and ramifies on the deep-seated muscles at the upper and inner part of the thigh.

2. *Veins.*

The FEMORAL VEIN adheres closely to the femoral artery in its passage out of the abdomen, and accompanies it in its course down the thigh, where it passes under Poupart's ligament; it lies on the inside of the artery, but, as it descends, it turns more and more posteriorly, so that where they perforate the tendon of the adductor magnus, the vein is situated fairly behind the artery.

Its branches correspond to those of the femoral artery; but about an inch below Poupart's ligament, it receives the vena saphena major, to which there is no corresponding artery.

The OBTURATOR VEIN accompanies the obturator artery, and has the same distribution.

3. *Nerves.*

Femoral nerve, NERVUS CRURALIS ANTERIOR, or the Anterior Crural nerve, where it passes from under Poupart's ligament, lies about half an inch on the outside of the femoral artery; it immediately divides into a number of branches, which supply the muscles and integuments on the forepart and outside of the thigh: a considerable branch, however, accompanies the femoral artery, leaves that vessel where it is about to perforate the adductor magnus, and appears as a cutaneous nerve on the inside of the knee; proceeding downward on the inside of the leg, it accompanies the saphena vein, and terminates on the inner ankle, and upper part of the foot. This branch is named NERVUS SAPHENUS LONGUS.

Another branch, which, running close to the femoral vessels, pierces the fascia, and is distributed below the knee. The SHORT SAPHENOUS.

MIDDLE CUTANEOUS to the sartorius and skin.

MUSCULAR to muscles principally on the outer side of the thigh.

NERVUS OBTURATOR, or the obturator nerve, is found accompanying the obturator artery and vein; it has the same distribution, and some branches extend as far as the internal condyle of the thigh-bone, and communicate with the nervus saphenus.

Dissection of the Forepart of the Leg and Foot.

Carry an incision from the tubercle of the tibia along its spine, over the dorsum of the foot to the toes, where it may be crossed by a second from the inside to the outside of the foot. Reflect the integuments. The skin must be dissected off the toes separately. The fascia covering the muscles on the front of the leg is very strong. It is fixed to the heads of the tibia and fibula, is strengthened by a contribution from the Fascia Lata Femoris, and by its deep surface gives origin, in part, to the muscles. At the ankle it becomes very strong, and adhering to the outer and inner malleolus, forms the ANNULAR LIGAMENT which binds down the tendons. Before removing this fascia, notice the SAPHENA INTERNA VEIN and NERVE, running along the inside of the leg, and over the ankle to the foot, and around the external malleolus. The SAPHENA externa vein and nerve to the outer part of the foot, and over the outer ankle filaments of the peroneal cutaneous nerve (from the Peroneal).

The fascia should then be dissected off, and, in doing this, remark that it sends down processes betwixt the muscles; these are named intermuscular ligaments; they give origin to the fibres of all the muscles betwixt which they pass, connecting them together inseparably, so that the dissection has a rough appearance.

Muscles situated on the Forepart and Outside of the Leg.

These are six in number.

1. The TIBIALIS ANTICUS—*Arises*, principally fleshy, from the exterior surface of the tibia, from its anterior angle or spine, and from nearly half of the interosseous ligament, and two-thirds of the length of the bone; also from the inner surface of the fascia of the leg, and from the intermuscular ligaments. The fleshy fibres descend obliquely, and terminate in a strong tendon, which crosses from the outside to the forepart of the tibia, passes through a distinct ring of the annular ligament near the inner ankle, runs over the astragalus and os naviculare, and is

Inserted into the upper and inner part of the os cuneiforme internum, and the base of the metatarsal bone supporting the great toe.

Situation. The belly is quite superficial, lying under the fascia of the leg on the outside of the spine of the tibia.

Use. To draw the foot upward and inward; or, in other words, to bend the ankle-joint.

2. EXTENSOR LONGUS DIGITORUM PEDIS—*Arises*, tendinous and fleshy, from the outer part of the head of the tibia; from the head of the fibula; from the anterior angle of the fibula almost its whole length, and from part of the smooth surface between the anterior and internal angles; from a small part of the interosseous ligament; from the fascia and intermuscular ligaments.

Below the middle of the leg, it splits into four round tendons, which pass under the annular ligament, become flattened, and are

Inserted into the root of the first phalanx of each of the four small toes, and expanded over the upper side of the toes as far as the root of the last phalanx.

Situation. This muscle also runs entirely superficial; it lies between the tibialis anticus and peroneus longus, but at the lower part of the leg it is separated from the tibialis anticus by the extensor pollicis longus.

Use. To extend all the joints of the four small toes: to bend the ankle-joint.

3. PERONEUS TERTIUS—*Arises*, fleshy, from the anterior angle of the fibula, and from part of the smooth surface between the anterior and internal angles, extending from below the middle of the bone downward to near its inferior extremity; sends its fleshy fibres forward to a tendon, which passes under the annular ligament, in the same sheath as the extensor digitorum longus, and is

Inserted into the base of the metatarsal bone of the little toe.

Situation. The belly is inseparably connected with the extensor longus digitorum, and is properly the outer part of it.

Use. To assist in bending the foot.

4. EXTENSOR PROPRIUS POLLICIS PEDIS — *Arises*, tendinous and fleshy, from part of the smooth surface between the anterior and internal angles of the fibula, and from the neighboring part of the interosseous ligament, extending from some distance below the head of the bone to near its inferior extremity; the fibres pass obliquely downward and forward into a tendon, which, inclining inward, passes over the forepart of the astragalus and os naviculare, to be

Inserted into the base of the first and of the second phalanges of the great toe.

Situation. The belly is concealed between the tibialis anticus and extensor digitorum longus, and cannot be seen till those muscles are separated from one another.

Use. To extend the great toe; and to bend the ankle.

The anterior tibial artery and nerve will be seen at the upper third of the limb between the extensor communis of the toes and the tibialis anticus, at the middle between the last and the extensor pollicis, and at the ankle between the last and the extensor communis digitorum.   Branches will be described presently.

5. The PERONEUS LONGUS—*Arises*, tendinous and fleshy, from the forepart and outside of the head of the fibula, and from the adjacent part of the tibia, from the

Fig. 61.

A Side View of the Muscles of the Leg and Foot.

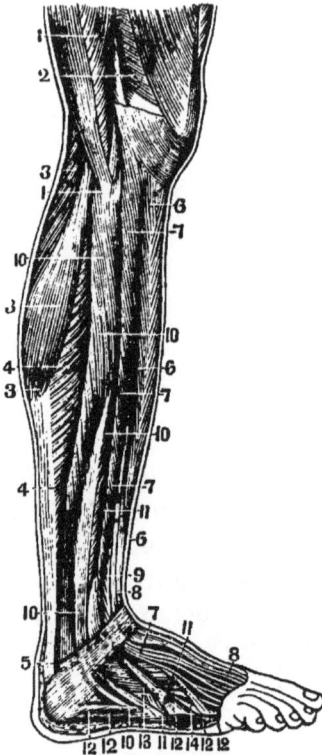

1. Biceps Flexor Cruris.
2. Vastus Externus.
3, 3. Gastrocnemius.
4. Soleus.
5. Tendo Achillis.
6. Tibialis Anticus.
7. Extensor Longus Digitorum Pedis.
8. Extensor Proprius Pol-
   licis.
9. Peroneus Tertius.
10. Peroneus Longus.
11. Peroneus Brevis.
12, 12. Abductor Minimi Digiti.
13. Extensor Brevis Digitorum.
14. Interosseus Dorsalis.

external angle of the fibula, and from the smooth surface between the anterior and external angles as far down as one-third of the length of the bone from its lower extremity; also from the fascia of the leg and intermuscular ligaments. The fibres run obliquely outward into a tendon, which passes behind the outer ankle, through a groove in the lower extremity of the fibula; is then reflected forward through a superficial fossa in the outside of the os calcis, passes over a projection, runs in a groove in the os cuboides, passes over the muscles in the sole of the foot, and is

Inserted, tendinous, into the outside of the base of the metatarsal bone of the great toe, and into the os cuneiforme internum.

Situation. The belly is quite superficial; it lies between the outer edge of the extensor longus digitorum and the anterior edge of the soleus. The tendon is superficial where it crosses the outside of the os calcis, but, in the sole of the foot, is concealed by the muscles situated there, and will be seen in the dissection of that part.

Use. To extend the ankle-joint, turning the sole of the foot outward.

6. The PERONEUS BREVIS—*Arises*, fleshy, from the outer edge of the anterior angle of the fibula, and from part of the smooth surface behind that angle; beginning about one-third down the bone, and continuing its adhesion to near the ankle; from the fascia of the leg, and from the intermuscular ligaments. The fibres run obliquely toward a tendon, which passes through a groove of the fibula behind the outer ankle, being there inclosed in the same ligament with the tendon of the peroneus longus, then through a separate groove on the outside of the os calcis, and is

Inserted into the external part of the base of the metatarsal bone that sustains the little toe.

Situation. This muscle arises between the extensor longus digitorum and peroneus longus; its belly is overlapped, and concealed by the belly of the peroneus longus.

Below it is separated from the peroneus tertius by that projection of the fibula which forms the outer ankle, and which is only covered by the common integuments.

Use. Same as that of the peroneus longus.

Muscles on the Upper Part of the Foot.

Only one muscle is found in this situation.

EXTENSOR BREVIS DIGITORUM PEDIS—*Arises*, fleshy and tendinous, from the anterior and upper part of the os calcis, from the os cuboides, and from the astragalus; forms a fleshy belly, divisible into four portions; these send off four slender tendons, which are

Inserted, the first tendon into the first phalanx of the great toe, and the other three into all the small toes except the little one, uniting with the tendons of the extensor digitorum longus, and being attached to the upper convex surface of all the phalanges.

Situation. The belly of this muscle lies under the tendons of the extensor digitorum longus and peroneus brevis.

Use. To extend the toes.

Of the Vessels and Nerves in the Forepart of the Leg and Foot.

1. *Arteries.*

ARTERIA TIBIALIS ANTICA.—The anterior tibial artery passes from the ham betwixt the inferior edge of the popliteus and the superior fibres of the soleus, and then through a large perforation in the interosseous ligament, to reach the forepart of the leg; this perforation is much larger than the size of the artery, and is filled up by the fibres of the musculus tibialis posticus, which may thus be said to arise from the forepart of the tibia. The artery then runs down close upon the middle of the interosseous ligament, between the tibialis anticus and extensor proprius pollicis; below the middle of the leg it leaves the interosseous ligament, and passes gradually more forward; it crosses under the tendon of the extensor proprius pollicis, and is then situated between that tendon and the first tendon of the extensor longus digitorum. At the ankle it runs over the forepart of the tibia, being now situated more superficially; then over the astragalus and os naviculare, and over the junction of the os cuneiforme internum and medium, crossing under that tendon of the extensor brevis digitorum which goes to the great toe. Arriving at the space between the bases of the two first metatarsal bones, it plunges into the sole of the foot, and immediately joins the plantar arch.

Branches.—1. A. RECURRENS, which ramifies over the forepart of the knee, inosculating with the articular arteries.

2. Numerous twigs to the tibialis anticus, extensor pollicis, and other muscles on the forepart of the leg.

3. A. MALLEOLARIS INTERNA ramifies over the inner ankle, and inosculates with the peroneal and posterior tibial arteries.

4. The EXTERNAL MALLEOLAR ramifies over the outer ankle.

Fig. 62.

A VIEW OF THE ANTERIOR TIBIAL ARTERY.

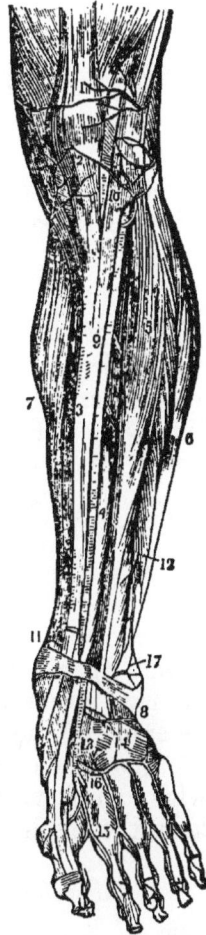

1. Tendon of the Rectus Muscle.
2. Ligament of the Patella.
3. Tibia.
4. Extensor Proprius Pollicis Pedis.
5. Extensor Communis Digitorum Pedis.
6. Peroneus Longus and Brevis Muscles.
7. Inner Border of the Gastrocnemius and Soleus Muscles.
8. Anterior Annular Ligament.
9. Anterior Tibial Artery.
10. Recurrent Articular Branch.
11. Internal Malleolar Branch.
12. Anterior Peroneal Artery.
13. Dorsal Artery of the Great Toe.
14. Tarsal and Metatarsal Branches.
15. Branch to the Great Toe.
16. Terminal Branch to join the Plantar Arch.
17. External Malleolar Artery.

5. The TARSAL and METATARSAL ARTERIES are two small branches which cross the tarsal and metatarsal bones, and pass obliquely to the outer edge of the foot.

From the tarsal or metatarsal artery come off the IN-

TEROSSEAL ARTERIES, which supply the interosseal
spaces and the back part of the toes.

6. A large branch comes off from the anterior tibial,
where it is about to plunge into the sole of the foot; it
runs along the space betwixt the two first metatarsal
bones, and at the anterior extremity of those bones bifurcates into—

(1) A. DORSALIS HALLUCIS, a considerable branch,
which runs on the back part of the great toe.

(2) A branch which runs on the inner edge of the
toe next to the great one.

2. *Veins.*

The ANTERIOR TIBIAL VEIN consists of two branches,
which accompany the artery and its ramifications.

3. *Nerves.*

The ANTERIOR TIBIAL NERVE is a branch of the peroneal nerve; it is seen in the ham arising from the peroneal and crossing under the muscles on the outside of the
fibula; it emerges from under the extensor longus digitorum, comes in contact with the anterior tibial artery,
and accompanies it down the leg; it is distributed on the
back part of the foot and toes.

Posterior Part of the Thigh.

Above that part of the fascia which invests the thigh
behind there are several cutaneous nerves. They
originate either from the lumbar nerves and come over
the spine of the ilium, or from the great sciatic nerve
emerging under the edge of the gluteus maximus. Other
twigs come from the sacral nerves or the sciatic as *it passes
down* the thigh.

Muscles on the Back Part of the Thigh.

There are eleven. Carry an incision along the crest
of the ilium to the spine, thence over the sacrum to the
point of the os coccygis. Reflect the integuments.

GLUTEUS MAXIMUS—Is usually covered by a large amount of adipose substance. It arises, fleshy, from the posterior third of the crest of the ilium, from the whole lateral surface of the sacrum below the posterior spinous process, from the back part of the inferior sacro-sciatic ligament, over which the muscle projects, and from the side of the os coccygis. The fibres converge to form a strong flat tendon, which glides over the trochanter major, is connected to the fascia lata, and is

Inserted into the rough surface on the outer part of the linea aspera immediately below the trochanter.

Situation. It is quite superficial, covering all the other muscles which are situated on the back part of the hip, covering also the tuber ischii, and the tendons of the muscles which arise from that projection. Its insertion lies between the vastus externus and the adductor magnus femoris.

Use. To restore the thigh, after it has been bent; to rotate it outward; to extend the pelvis on the thigh, and maintain it in that position in the erect posture of the body.

The muscle is now to be lifted from its origin, and left hanging by its tendon; remark the large bursa mucosa formed between the tendon and the trochanter major.

2. The GLUTEUS MEDIUS—*Arises*, fleshy, from all the outer edge of the spine of the os ilium, as far as the posterior tuberosity; from the dorsum of the bone, between the spine, and semicircular ridge; also from the rough surface which extends from the anterior superior to the anterior inferior spinous process, and from the inside of a fascia which covers its anterior part. The fibres converge into a strong and broad tendon, which is

Inserted into the upper and outer part of the great trochanter.

Situation. The posterior part of the belly and the tendon are concealed by the gluteus maximus, but the anterior and largest part of this muscle is superficial, being covered by a strong fascia.

Use. To draw the thigh-bone outward, or away from

the opposite limb; to maintain the pelvis in a state of equilibrium on the thigh in progression, while the other foot is raised from the ground; by its posterior fibres to rotate the limb outward; and by its anterior inward.

Having lifted up this muscle from its origin, you will discover

Fig. 63.

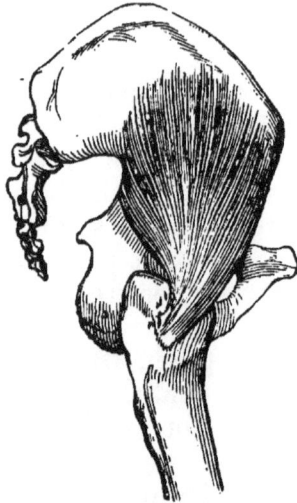

A VIEW OF THE GLUTEUS MINIMUS MUSCLE.

3. The GLUTEUS MINIMUS.—It *arises*, fleshy, from the semicircular ridge of the ilium, and from the dorsum of the bone below the ridge within half an inch of the acetabulum. Its fibres run in a radial direction toward a strong tendon, which is

Inserted into the anterior and superior part of the great trochanter.

Situation. It is entirely concealed by the gluteus medius.

Use. Same as that of the preceding.

4. The PYRIFORMIS—*Arises*, within the pelvis, by three tendinous and fleshy origins, from the second, third, and fourth false vertebræ or divisions of the sacrum. It forms a thick belly, which passes out of the pelvis through the great sacro-ischiatic foramen above the superior sacro-sciatic ligament.

Inserted, by a roundish tendon, into the uppermost part of the cavity of the root of the trochanter major.

Situation. Like the other small muscles of the hip, it is entirely concealed by the gluteus maximus; its belly lies behind and below the gluteus medius.

Use. To move the thigh a little upward, and roll it outward. The great sciatic nerve, the ischiatic and internal pudic bloodvessels and nerves, pass out of the pelvis below this muscle, and the gluteal bloodvessels and nerve above it.

5. The GEMINI consist of two heads, which are distinct muscles.

(1) The superior arises from the back part of the spinous process of the ischium.

(2) The inferior from the upper part of the tuberosity of the os ischium, and the anterior surface of the posterior sacro-sciatic ligament.

Inserted, tendinous and fleshy, into the cavity at the root of the trochanter major, immediately below the insertion of the pyriformis, and above the insertion of the obturator externus.

Situation. Like the other muscles, they are covered by the gluteus maximus; they lie below the pyriformis, and above the quadratus femoris.

Use. To roll the thigh outward, and to bind down the tendon of the obturator internus.

Lying between the bellies of the gemini, you will perceive

6. The OBTURATOR INTERNUS.—It *arises*, tendinous and fleshy, from more than one-half of the internal circumference of the foramen thyroideum, and from the inner surface of the ligament which fills up that hole; it forms a flattened tendon, which passes out of the pelvis in a sinuosity betwixt the spinous process and tuberosity of the ischium, and, becoming rounder, is

Inserted into the pit at the root of the trochanter major.

Situation. Its origin lies within the pelvis, and cannot be exposed till the contents of that cavity are removed; the tendon, where it passes through the notch

in the ischium, is seen projecting between the two
origins of the gemelli. There is a bursa mucosa be-

Fig. 64.

A VIEW OF THE MUSCLES ON THE BACK OF THE HIP.

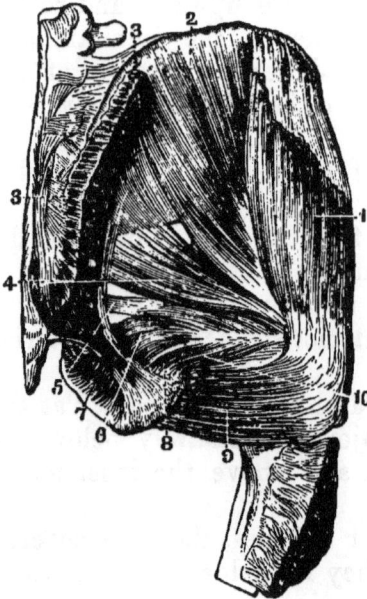

1, 2. Gluteus Medius.
3. Cut Origin of Gluteus
Maximus.
4. Pyriformis.
5, 8, 10. Gemelli.
6, 7. Obturator Internus.
2. Quadratus Femoris.

twixt the tendon of this muscle and the surface of the
ischium over which it glides.

Use. To roll the os femoris obliquely outward.

7. The QUADRATUS FEMORIS—*Arises*, tendinous and
fleshy, from an oblique ridge, which descends from the
inferior edge of the acetabulum along the body of the
ischium, between its tuberosity and the foramen thyroideum; its fibres run transversely, to be

Inserted, fleshy, into a rough ridge on the back part
of the femur, extending from the root of the greater
trochanter to the root of the lesser.

Situation. It is concealed by the gluteus maximus; its
origin is in contact with the origin of the hamstring
muscles.

Use. To roll the thigh outward.

On lifting up the quadratus femoris from its origin, and leaving it suspended by its insertion, you discover, running in the same direction, the strong tendon of

8. The OBTURATOR EXTERNUS.—This muscle *arises*, fleshy, from almost the whole circumference of the foramen thyroideum, and from the external surface of the obturator ligament; its fibres pass outward through the notch placed between the inferior margin of the acetabulum and the tuberosity of the ischium, wind around the cervix of the os femoris, adhering to the capsular ligament, and terminate in a strong tendon, which is

Inserted into the lowermost part of the cavity, at the root of the trochanter major, immediately below the insertion of the inferior head of the gemini.

Situation. This muscle cannot be distinctly seen, until all the muscles which run from the pelvis to the upper part of the thigh are removed, both on the fore and back part.

Use. To roll the thigh-bone obliquely outward.

9. The BICEPS FLEXOR CRURIS—*Arises* by two distinct heads: the first, called the LONG HEAD, arises in common with the semitendinosus, by a short tendon, from the outer part of the tuberosity of the ischium, and, descending, forms a thick fleshy belly. The second, termed the SHORT HEAD, arises, tendinous and fleshy, from the linea aspera, immediately below the insertion of the gluteus maximus; and from the oblique ridge running to the outer condyle, where it is connected with the fibres of the vastus externus. The two heads unite at an acute angle, a little above the external condyle, and terminate in a strong tendon, which is

Inserted into a rough surface on the outside of the head of the fibula.

Situation. The long head of this muscle is concealed at its upper part by the inferior fibres of the gluteus maximus; below this, it is situated quite superficial,— it forms the outer hamstring.

Use. To bend the leg, and particularly by means of its shorter head to twist the leg outward in the bent state of the knee.

10. The SEMITENDINOSUS—*Arises,* tendinous, in common with the long head of the biceps, from the tuberosity of the ischium; it has also some fleshy fibres arising from that projection more outwardly: as it descends, it arises for two or three inches, fleshy, from the inside of the tendon of the biceps; forms a thick belly, and term-

Fig. 65.

A VIEW OF THE PRINCIPAL MUSCLES OF THE BACK OF THE THIGH.

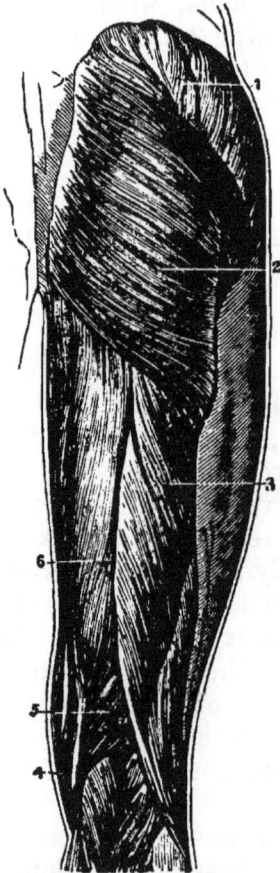

1. Gluteus Medius.
2. Gluteus Maximus.
3. Biceps Flexor Cruris.
4. Tendon of Semitendinosus.
5. Semimembranosus Muscle.
6. Semitendinosus Muscle.

inates at the distance of three or four inches from the knee in a long round tendon, which, becoming flat, passes behind the head of the tibia, and is reflected forward, to be

Inserted into the anterior angle of that bone, some little way below its tubercle.

Situation. This muscle, as well as the biceps, is covered above by the gluteus maximus ; its belly lies between the biceps flexor and gracilis, and is situated entirely superficial.

Use. To bend the leg backward, and a little inward.

11. The SEMIMEMBRANOSUS — *Arises*, by a strong round tendon, from the upper and outer part of the tuberosity of the ischium ; the tendon, soon becoming broader, sends off obliquely a fleshy belly ; this muscle is continued, fleshy, much lower down than that last described. The fleshy fibres terminate obliquely in another flat tendon, which passes behind the inner condyle, sends off a thin aponeurotic expansion under the inner head of the gastrocnemius, to cover the posterior part of the capsule of the knee-joint, and to be affixed to the external condyle : the tendon then becoming rounder, is

Inserted into the inner and back part of the head of the tibia.

Situation. This is a semipenniform muscle ; its origin lies anterior to the tendinous origin of the two last muscles, and more outwardly, being situated between them and the origin of the quadratus femoris.

Use. To bend the leg backward.

The two last described muscles properly form the inner hamstring ; but some enumerate among the tendons of the inner hamstring, the sartorius and gracilis.

Vessels and Nerves on the Posterior Part of the Thigh.

Arteries.

1. ARTERIA GLUTEA, or ILIACA POSTERIOR.—This is the largest branch of the internal iliac artery ; it passes out of the pelvis at the upper part of the sciatic notch. On raising the gluteus maximus, and medius, this artery is seen coming over the pyriformis, betwixt the superior edge of that muscle and the inferior edge of the os ilium. The principal trunk passes under the gluteus medius, and ramifies on the dorsum of the os ilium ; other large

branches are also continued to the gluteus maximus, and the muscles situated on the back part of the pelvis.

2. ARTERIA SCIATICA, vel ISCHIATICA, is another large branch of the internal iliac, which comes out from under the pyriformis, betwixt the lower edge of that muscle and the superior sacro-sciatic ligament; its principal branches descend between the trochanter major and tuberosity of the ischium; it sends other twigs round toward the anus and perineum.

Both these arteries inosculate with the other branches of the internal and external iliac.

The Internal Pudic, a branch of the ischiatic, within the pelvis, comes out below the pyriformis, and re-enters the pelvis through the lesser ischiatic opening, and continuing up along the ramus of the ischium and pubes, is ultimately distributed to the penis.

The VEINS correspond exactly to the arteries. They terminate in the internal iliac vein.

Nerves.

NERVUS SCIATICUS, vel ISCHIATICUS, or the Great Sciatic Nerve, from the sacral plexus, is seen coming out of the pelvis, below the pyriformis. It descends over the gemini and quadratus femoris in the hollow betwixt the great trochanter and the tuberosity of the ischium—runs down the back part of the thigh, anterior to, i.e. nearer the bone than the hamstring muscles; being situated between the anterior surface of the semi-membranosus and the posterior surface of the triceps adductor longus. After sending off the peroneal nerve, it arrives in the ham, and becomes the POPLITEAL NERVE. In this course it gives off several branches to the muscles and integuments. It sometimes perforates the belly of the pyriformis by distinct trunks, which afterward unite.

Gluteal Nerve, derived from the fifth lumbar nerve, and accompanying the Gluteal Artery.

INTERNAL PUDIC NERVE, derived from the sacral plexus, and follows the course of Internal Pudic Artery and its branches.

Lesser Ischiatic—from the sacral plexus : passes out below the pyriformis, and passes down the thigh to the ham, where it becomes connected with the external saphenous nerve. Its branches are muscular and cutaneous; to the gluteus maximus integuments of the hip, etc. One branch, the *Inferior Pudendal*, curves around the tuber ischii, and is distributed to the scrotum.

Dissection of the Ham and Fascia on the Back Part of the Thigh.

On removing the integuments from the back part of the knee-joint and leg, we observe a Fascia, which covers the great vessels and muscles. It is evidently continued from the great fascia of the thigh, is strengthened by adhesions to the condyles of the femur, and to the head of the fibula, and is prolonged upon the muscles on the back of the leg.

Upon dissecting back that part of the fascia which covers the ham, the GREAT SCIATIC NERVE appears lying between the outer and inner hamstring muscles. This nerve, having given off branches about the ham, and to the integuments on the back of the leg, divides at some distance above the condyles of the femur into two large branches.

1. The greater Nerve continues its course betwixt the heads of the gastrocnemii muscles. In the ham it is named the POPLITEAL NERVE, and where it descends in the leg, POSTERIOR TIBIAL.

2. The lesser Nerve, which is the external branch, is named the PERONEAL or FIBULAR NERVE; it passes outward and obliquely downward, runs between the external head of the gastrocnemius, and the tendon of the biceps flexor cruris; and sinks among the muscles which surround the head of the fibula.

Branches of the Peroneal Nerve.

(1) Cutaneous branches are sent off from the peroneal nerve at its uppermost part over the gastrocnemius to the integuments in the back part of the leg, and outer

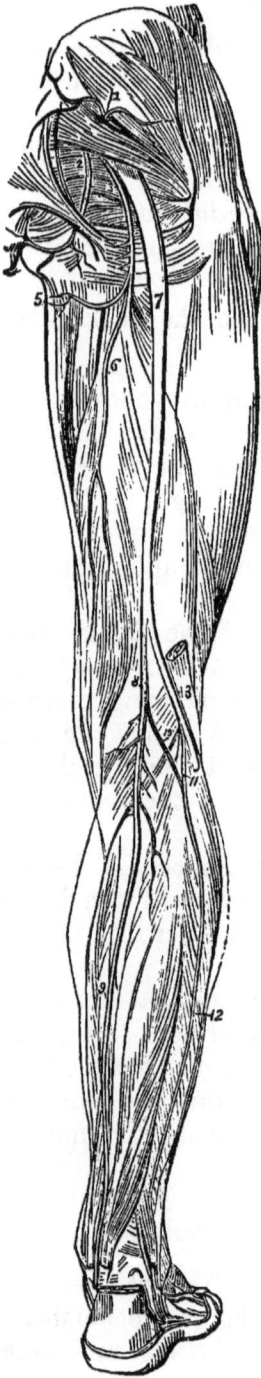

Fig. 66.
A VIEW of THE GREATER AND LESSER
SCIATIC NERVES IN THEIR ENTIRE
COURSE DOWN THE LIMB.

1. Superior Gluteal Nerve.
2. Pudic Nerves.
3. Lesser Sciatic Nerve.
5. Inferior Pudendal Branch.
6. Continuation of the Small Sciatic.
7. Greater Sciatic Nerve.
8, 9. Popliteal, and Posterior Tibial Nerve.
10, 12. Short Saphenous Nerve.
11. Peroneal Communicating Branch.
13. Peroneal Nerve.

side of the foot. Branches also are distributed about the joint.

(2) A large branch, the ANTERIOR TIBIAL NERVE, passes under the flesh of the peroneus longus and extensor longus digitorum, where those muscles arise from the heads of the fibula and tibia; and comes in contact with the anterior tibial artery, which it accompanies down the leg.

(3) Another branch passes into the upper extremity of the peroneus longus, and is continued in the substance of that muscle for some space. It then emerges from beneath it, and continues its course under the muscles on the forepart and outside of the leg; it pierces the fascia, and, becoming cutaneous, is lost on the ankle and upper surface of the foot. (The PERONEAL CUTANEOUS.)

Below the great sciatic nerve, there is much cellular membrane and fat, which being removed, the GREAT POPLITEAL VEIN is exposed. It adheres to the POPLITEAL ARTERY, which lies under it close upon the bone.

ARTERIA POPLITEA is the trunk of the FEMORAL, which assumes that name, after it has perforated the tendon of the adductor magnus. It lies between the condyles of the femur, close upon the bone, and descends between the heads of the gastrocnemius. At the lower edge of the popliteus, the popliteal artery divides into the ANTERIOR and POSTERIOR TIBIAL ARTERIES.

BRANCHES.—1. ARTICULAR ARTERIES are four or five small twigs, ramifying over the knee-joint and neighboring muscles, inosculating with one another, and with the arteries below the knee.

2. Two branches are sent to the two heads of the gastrocnemii muscles; the sural twigs are also given to the soleus, plantaris, etc.

The POPLITEAL VEIN receives branches corresponding to those of the artery; it lies behind the artery in the erect posture.

About two inches above the condyle, it receives the VENA SAPHENA MINOR, which returns the blood from the outer side of the foot; the trunk of this vein lies under the fascia on the back of the leg.

Dissection of the Posterior Part of the Leg.

The fascia investing the posterior part of the leg is much thinner than on the front part. Remove it and the muscles on this aspect of the limb may be exposed.

They are seven in number.

1. GASTROCNEMIUS—*Arises*, by two heads, from the upper and back part of the condyles of the os femoris. Each head forms a fleshy belly, the fibres of which are oblique, passing from a tendinous expansion which covers the posterior surface of the muscle to another which covers the anterior surface. The internal belly is the largest, and they are separated by a triangular interval, in which the popliteal bloodvessels and nerves pass to the leg. These heads unite a little below the knee, in a middle tendinous line, and below the middle of the tibia ends in a strong flat tendon which joins that of the soleus. In the groove on its surface lies the saphena externa nerve and vein.

Reflect the two heads of the gastrocnemius from the femoral condyles, and you will then expose

2. The SOLEUS, or GASTROCNEMIUS INTERNUS—which *arises* by two origins or heads. The first, or External Origin, which is by much the largest, arises, principally fleshy, from the posterior surface of the head of the fibula, and from the external angle of that bone, for two-thirds of its length, immediately behind the peroneus longus. The second, or internal head, arises, fleshy, from an oblique ridge on the posterior surface of the tibia, just below the popliteus, and from the inner angle of that bone, during the middle third of its length. The two heads, which are separated at first by the posterior tibial artery and nerve, unite immediately, form a large belly, which, covered by the tendon of the gastrocnemius, is continued, fleshy, to within a short distance of the ankle-joint; a little above which the tendons of the gastrocnemius and soleus unite, and form a strong round tendon, named the

TENDON ACHILLIS, which slides over the upper and posterior part of the os calcis, where it is furnished with a small bursa mucosa, to be

Inserted into a rough surface on the back part of that bone.

Situation. The gastrocnemius arises between the hamstring tendons. Its belly is superficial, and forms the upper or greater calf of the leg.

Fig. 67.

THE SUPERFICIAL MUSCLES OF THE POSTERIOR FACE OF THE LEG.

1. The Biceps Muscle forming the Outer Hamstring.
2. The Tendons forming the Inner Hamstring.
3. The Popliteal Space.
4. The Gastrocnemius Muscle.
5. The Soleus.
6. Tendo Achillis.
7. The Posterior Tuberosity of the Os Calcis.
8. The Tendons of the Peroneus Longus and Brevis Muscles passing behind the Outer Ankle.
9. The Tendons of the Tibialis Posticus and Flexor Longus Digitorum Pedis passing into the Foot behind the Inner Ankle.

The soleus has its largest part concealed by the gastrocnemius, but part of it appears on each side of the belly of that muscle. There is a bursa mucosa betwixt the upper part of the os calcis and the tendo Achillis.

Use. To elevate the os calcis, and thereby to lift up the whole body as a preparatory measure to its being carried forward in progression; to carry the leg backward on the foot when that is fixed; the gastrocnemius, from its origin in the thigh, also bends the leg on the thigh.

The heads of the gastrocnemius should now be lifted up, which will expose

3. The PLANTARIS.—This muscle *arises*, fleshy, from the upper part of the external condyle, and from the oblique ridge above that condyle, forms a pyramidal belly about three inches in length, which adheres to the capsule of the knee-joint, runs over the popliteus, and terminates in a long, slender, thin tendon. This tendon passes obliquely inward over the inner head of the soleus, and under the gastrocnemius; emerges from between those two muscles, where their tendons unite, and then runs down by the inside of the tendo Achillis, to be

Inserted into the posterior part of the os calcis, on the inside of the insertion of the tendo Achillis, and somewhat before it.

Situation. The origin and belly of this muscle are concealed by the external head of the gastrocnemius.

Use. To extend the foot and roll it inward, and to assist in bending the leg.

4. The POPLITEUS—*Arises*, within the capsular ligament of the knee, by a round tendon, from a deep pit or hollow on the outer side of the external condyle; adheres to the posterior and outer surface of the external semilunar cartilage; passes, within the cavity of the joint, over the side of the condyle to its back part; perforates the capsular ligament, and forms a fleshy belly, which runs obliquely inward, being covered by a thin tendinous fascia, to be

Inserted, broad, thin, and fleshy, into an oblique ridge on the posterior surface of the tibia, a little below its head, and into the triangular space above that ridge.

Situation. This muscle is concealed entirely by the gastrocnemius.

Use. To bend the leg, and, when bent, to roll it, so as to turn the toes inward.

The belly of the soleus should now be lifted, in order to expose the deeply-seated muscles. A strong membranous fascia (*Intermuscular*) is seen connecting and investing them, which is to be removed. This fascia also covers the posterior tibial vessels and nerve, but the description of the course of these vessels, though seen in this stage of the dissection, must be deferred.

The deep-seated muscles are:

The Flexor Longus Digitorum Pedis, situated behind the tibia.

The Flexor Longus Pollicis Pedis, situated behind the fibula.

The Tibialis Posticus, which is almost concealed by the two other muscles, and by the fascia, which connects them, and binds them down.

5. The FLEXOR LONGUS DIGITORUM PEDIS PERFORANS—*Arises*, fleshy, from the posterior flattened surface of the tibia, between its internal and external angles, below the attachment of the soleus, and continues to arise from the bone to within two or three inches of the ankle; the fibres pass obliquely into a tendon. This tendon runs behind the inner ankle in a groove of the tibia, passes under a strong ligament which goes from the inner ankle to the os calcis, and having received a strong tendinous slip from the flexor pollicis longus, divides about the middle of the sole of the foot into four tendons, which pass through the slits in the tendon of the flexor digitorum brevis, and are

Inserted into the extremity of the last joint of the four lesser toes.

The situation of the tendon is described with the muscles of the foot.

Use. To bend the last joint of the toes, and to assist in extending the foot.

6. FLEXOR LONGUS POLLICIS PEDIS—*Arises*, fleshy, from the posterior flat surface of the fibula, continuing its origin from some distance below the head of the bone to within an inch of the ankle. The fleshy fibres terminate in a tendon, which passes behind the inner ankle through a groove in the tibia; next through a groove in

10*

the astragalus, crosses in the sole of the foot the tendon of the flexor longus digitorum, to which it gives a slip of tendon; passes between the two sesamoid bones, and is

Inserted into the last joint of the great toe.

Situation. It lies on the outside of the flexor longus digitorum, between that muscle and the peroneus longus; the tendon will be seen in the foot.

Use. To bend the last joint of the great toe, and, being connected by a cross slip to the flexor digitorum communis, to assist in bending the other toes.

Fig. 68.
DEEP MUSCLES ON THE BACK OF THE LEG.

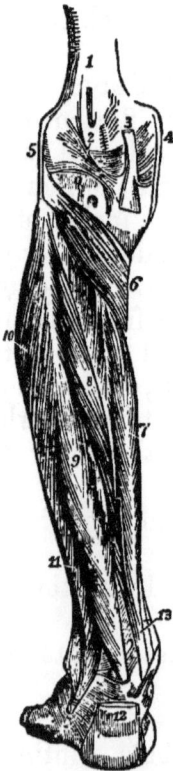

1. The Lower Extremity of the Femur.
2. Ligament of Winslow.
3. Tendon of the Semimembranous Muscle.
4. Internal Lateral Ligament of the Knee-joint.
5. External Lateral Ligament.
6. Popliteus Muscle.
7. Flexor Longus Digitorum Pedis.
8. Tibialis Posticus Muscle.
9. Flexor Longus Proprius Pollicis Pedis.
10. Peroneus Longus Muscle.
11. Peroneus Brevis.
12. Tendo Achillis divided near its Insertion.
13. Tendons of the Tibialis Posticus and Flexor Longus Digitorum Pedis, just as they are about to pass beneath the Internal Annular Ligament. The interval between the latter Tendon and the Tendon of the Flexor Longus Pollicis is occupied by the Posterior Tibial Vessels and Nerves.

7. The TIBIALIS POSTICUS—*Arises*, fleshy, from the posterior surface of both the tibia and fibula, immedi-

ately below the upper articulation of these bones with each other; from the whole of the interosseus ligament; from the angles of the bones to which that ligament is attached; and from the flat surface of the fibula behind its internal angle for more than two-thirds of its length. The fibres run obliquely toward a middle tendon, which, becoming round, passes behind the inner ankle through a groove in the tibia.

Inserted into the upper and inner part of the os naviculare, being further continued through a groove in that bone to the internal and external cuneiform bones.

Situation. The belly is concealed at its lower part by the flexor longus digitorum and flexor pollicis, and cannot be seen till those muscles are separated. The tendon crosses under that of the flexor longus digitorum above the ankle, and, where it passes through the groove in the tibia, is situated more forward than the tendon of that muscle.

Use. To extend the foot, and turn it inward.

Vessels and Nerves of the Posterior Part of the Leg.

1. *Arteries.*

ARTERIA TIBIALIS POSTICA. — The posterior tibial artery, which is the continued trunk of the popliteal, sinks under the origins of the soleus, and runs down the leg between that muscle and the more deeply scattered flexors of the toes. It does not lie in immediate contact with the fibres of the flexors, but, like the femoral artery, is invested by a strong sheath of condensed cellular membrane. It is, together with its veins and accompanying nerve, also supported by the fascia which binds down the deep-seated muscles. As it descends, it gradually advances more forward, following the course of the flexor tendons: it passes behind the inner ankle, lying posterior to the tendon of the flexor longus digitorum, and anterior to that of the flexor longus pollicis. It sinks under the abductor pollicis, arising from the os calcis, and immediately divides into two branches:

(1) The Internal Plantar Artery is the smallest, and

ramifies among the mass of muscles situated on the inner edge of the sole of the foot.

(2) The External Plantar Artery directs its course outward, and having reached the metatarsal bone of the little toe, forms the PLANTAR ARCH, which crosses the three middle metatarsal bones obliquely, about their

Fig. 69.
ARTERIES ON THE BACK OF THE THIGH, LEG, AND FOOT.

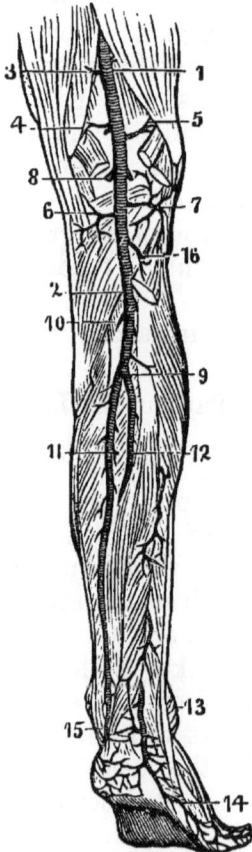

1, 2. Popliteal Artery.
3. Anastomotic Artery, the Last Branch of the Femoral.
4. Superior Internal Articular Artery.
5. Superior External Articular Artery.
6. Inferior Internal Articular.
7. Azygos Artery.
8. Sural or Gastrocnemial Arteries.
2. Point at which the Popliteal divides into the Anterior and Posterior Tibial Arteries.
9. Point at which the Posterior Tibial gives off the Peroneal Artery; being called thus far the Tibio-Peroneal Artery.
10. Nutritious Artery of the Tibia.
11. Continued Trunk of the Posterior Tibial Artery.
12. Peroneal Artery.
13. External Malleolar Artery.
14. External Plantar Artery.
15. Internal Malleolar Artery.
16. Inferior External Articular Artery.

middle, and terminates at the space betwixt the two first metatarsal bones, where the trunk of the anterior tibial artery joins the arch. The convexity of this arch is toward the toes, and sends off the following branches:

a. A small branch to the outside of the little toe.

b. Ramus digitalis primus, or the first digital artery, which runs along the space between the two last metatarsal bones, and bifurcates into two branches, one to the inner side of the little toe, and the other to the outer side of the next toe.

c. The second digital artery, which runs along the next interosseous space, and bifurcates in a similar manner.

d. The third digital artery.

e. The fourth, or GREAT DIGITAL ARTERY, which supplies the great toe, and the inner side of the toe next to it.

The concavity of the arch sends off the INTEROSSEAL arteries, three or four small twigs, which go to the deep-seated parts in the sole of the foot, and, PERFORATING between the metatarsal bones, inosculate with the superior interosseal arteries on the upper side of the foot.

The branches of the Posterior Tibial Artery in the leg are

1. The PERONEAL ARTERY, which comes off from the tibial a little after it has sent off the anterior tibial; it is generally of a considerable size; it runs upon the inside of the fibula, giving numerous branches to the peroneal muscles and flexor of the great toe; its course is irregular. At the lower part of the leg it splits into

a. A. PERONEA ANTERIOR, which passes betwixt the lower heads of the tibia and fibula, to the forepart of the ankle, where it is lost.

b. A. PERONEA POSTERIOR is properly the termination of the artery; it descends along the sinuosity of the os calcis, inosculating with the branches of the tibialis postica, and terminates in the posterior part of the sole of the foot.

2. Muscular branches arise from the artery as it descends; twigs also are sent over the heel and ankle.

2. *Veins.*

VENÆ TIBIALES POSTICÆ.—The posterior tibial veins are generally two in number; they accompany the artery, and terminate in the popliteal vein; they are formed of branches, which correspond to those of the artery.

3. *Nerves.*

The POSTERIOR TIBIAL NERVE, which is the continuation of the great sciatic nerve, sinks below the soleus, and accompanies the posterior tibial artery; it gives off numerous filaments to the muscles in its neighborhood. At first it is immediately behind the artery, gradually getting on the outside of it as it descends; so that where they pass along the sinuosity of the os calcis, the nerve is situated close in contact with the side of the artery, but nearer to the projection of the heel than that vessel is.   With the artery, it divides into

1. The internal plantar nerve, and
2. The external plantar nerve.—These nerves supply the muscles and integuments in the sole of the foot.

Dissection of the Sole of the Foot.

The cuticle is very much thickened on the sole of the foot from constant pressure; betwixt the integuments and plantar aponeurosis, we find a tough granulated fat, which adheres firmly to the aponeurosis, and is dissected off with difficulty.

APONEUROSIS, seu FASCIA PLANTARIS, is a very strong tendinous expansion, which arises from the projecting extremity of the os calcis, and passes to the root of the toes, covering and supporting the muscles of the sole of the foot.   Where it arises from the heel, it is thick, but narrow; as it runs over the foot it becomes broader and thinner; and it is fixed to the head of each of the metatarsal bones by a bifurcated extremity, which, by its splitting, leaves room for the tendons, etc. to pass.   It seems divided into three portions, which are connected by strong fasciculi of tendinous fibres; and fibres are sent down, forming perpendicular partitions among the muscles, and separating them into three classes:

1. The middle portion, which is the largest, and under which are contained the flexor brevis digitorum, and the tendons of the flexor longus and lumbricales.

2. The external lateral portion, which covers the muscles of the little toe.

3. The internal lateral portion, concealing the muscles of the great toe.

On removing the plantar aponeurosis, the first order of muscles in the sole of the foot is exposed; it consists of three muscles:

1. ABDUCTOR POLLICIS PEDIS—*Arises*, tendinous and fleshy, from the lower and inner part of the os calcis; from a ligament which extends from the os calcis to the os naviculare; from the inside of the os naviculare and cuneiforme internum; and from the fascia plantaris.

Inserted, tendinous, into the internal sesamoid bone and base of the first phalanx of the great toe.

Use. To move the great toe from the rest.

Fig. 70.

MUSCLES OF THE SIDE OF THE FOOT.

1. Abductor Pollicis.
2, 2. Its Tendon.
3, 3. Flexor Brevis Pollicis.
   4. Tendon of Flexor Longus Pollicis.
   5. Aponeurosis Plantaris, divided.
6, 7. Flexor Brevis Digitorum Pedis.
   7. Lumbricales.
   8. Abductor Minimi Digiti.
   9. Flexor Brevis Minimi Digiti.
 10. Interossei.

2. ABDUCTOR MINIMI DIGITI PEDIS—*Arises*, tendinous and fleshy, from the outer side of the os calcis, and

from a strong ligament which passes from the os calcis to the metatarsal bone of the little toe; also from the fascia plantaris.

Inserted, tendinous, into the base of the metatarsal bone of the little toe, and into the outside of the base of the first phalanx.

This muscle can frequently be divided distinctly into two portions.

Use. To move the little toes from the other toes.

3. FLEXOR BREVIS DIGITORUM PEDIS PERFORATUS— *Arises*, fleshy, from the anterior and inferior part of the protuberance of the os calcis, and from the inner surface of the fascia plantaris; also from the tendinous partitions betwixt it and the abductors of the great and little toe. It forms a thick fleshy belly, and sends off four tendons, which split for the passage of the tendons of the flexor longus digitorum, and are

Inserted into the second phalanx of the four lesser toes.

The tendon of the little toe is often wanting.

Use. To bend the second joint of the toes.

The first order of muscles being removed, or being lifted from their origins and left hanging by their tendons, the second order is exposed.

1. The tendon of the flexor longus digitorum pedis is seen coming from the inside of the os calcis, and, having reached the middle of the foot, dividing into its four tendons, which pass through the slits of the tendons of the flexor digitorum brevis, and are inserted into the base of the last phalanx of the four lesser toes.

2. The tendon of the Flexor longus pollicis is seen crossing under the tendon of the flexor longus digitorum, and, having given to it a short slip of tendon, proceeding between the two sesamoid bones to the base of the last phalanx of the great toe.

3. FLEXOR DIGITORUM ACCESSORIUS, or Massa Carnea JACOBII SYLVII—*Arises*, fleshy, from the sinuosity at the inside of the os calcis, and tendinous from that bone more outwardly. It forms a belly of a square form.

Inserted into the outside of the tendon of the flexor digitorum longus, just at its division.

Use. To assist the flexor longus.

4. LUMBRICALES PEDIS—*Arises*, by four tendinous and fleshy beginnings, from the tendons of the flexor longus digitorum, immediately after their division.

Inserted, by four slender tendons, into the inside of the first phalanx of the four lesser toes, and into the tendinous expansion that is sent from the extensors to cover the upper part of the toes.

Use. To promote the flexion of the toes, and to draw them inward.

The second order of muscles being removed, we expose the third order:

1. FLEXOR BREVIS POLLICIS PEDIS—It *arises*, tendinous, from the under and fore part of the os calcis, where it joins with the os cuboides; also from the os cuneiforme externum. It forms a fleshy belly, which is connected inseparably to the abductor and adductor pollicis.

Inserted, by two tendons, into the external and internal sesamoid bones; and it is continued on into the base of the first phalanx of the great toe.

Use. To bend the first joint of the great toe.

2. ADDUCTOR POLLICIS PEDIS—*Arises*, tendinous and fleshy, from a strong ligament which extends from the os calcis to the os cuboides, and from the roots of the second, third, and fourth metatarsal bones. It forms a fleshy belly, which seems at its beginning divided into two portions.

Inserted, tendinous, into the external sesamoid bone and root of the metatarsal bone of the great toe.

Use. To bring this toe nearer the rest.

3. FLEXOR BREVIS MINIMI DIGITI PEDIS—*Arises*, tendinous and fleshy, from the os cuboides, and from the root of the metatarsal bone of the little toe.

Inserted, tendinous, into the base of the first phalanx of the little toe and into the anterior extremity of the metatarsal bone.

Use. To bend this toe.

4. TRANSVERSALIS PEDIS—*Arises*, tendinous, from the anterior extremity of the metatarsal bone supporting

the little toe; becoming fleshy, it crosses over the anterior extremities of the other metatarsal bones.

Fig. 71.

DISSECTION OF A SECOND LAYER OF THE PLANTAR MUSCLE OF THE
FOOT.

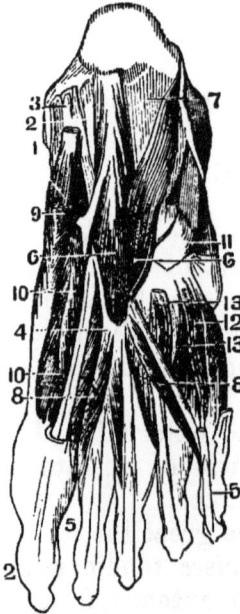

1. Tendon of Tibialis Posticus.
2. Tendon of Flexor Longus Pollicis.
3. Tendon of Flexor Longus Digitorum.
4. Point where it separates into four Tendons.
5. Points of Insertion.
6. Flexor Accessorius.
7. Calcaneo-cuboid Ligament.
8. Lumbricales Pedis.
9. Adductor Pollicis.
10. Flexor Brevis Pollicis.
11. Tendon of Peroneus Longus.
12. Flexor Brevis Minimi Digiti.
13. Interossei Muscles.

·*Inserted*, tendinous, into the anterior extremity of the metatarsal bone of the great toe, and into the internal sesamoid bone adhering to the adductor pollicis.

Use. To contract the foot, by bringing the toes nearer each other.

Ranging with this order of muscles, we may also observe—

A broad strong ligament, passing from the anterior sinuosity of the os calcis over the surface of the os cuboides.

The tendon of the tibialis posticus, dividing into numerous tendinous slips, to be inserted into the bones of the tarsus.

Having removed the muscles last described, we expose the fourth and last order.

The tendon of the peroneus longus is seen passing along a groove in the os cuboides, and crossing the tarsal bones, to be inserted into the base of the metatarsal bone of the great toe, and into the internal cuneiforme and second metatarsal bones.

Fig. 72.

PLANTAR INTEROSSEI.

1. Metatarsal Bone of the Great Toe.
2, 2, 2. Interosseous Muscles.
3, 3, 3. Their Insertion into the First Phalanx.

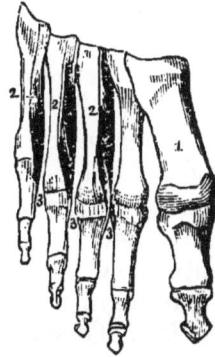

INTEROSSEI PEDIS INTERNI are three in number, situated in the sole of the foot. They *arise*, tendinous

Fig. 73.

DORSAL INTEROSSEI MUSCLES.

1. First Metatarsal Bone.
2, 2. Interossei Muscles.
3, 3, 3, 3. Their tendinous insertion into the first Phalanges of the Toes.

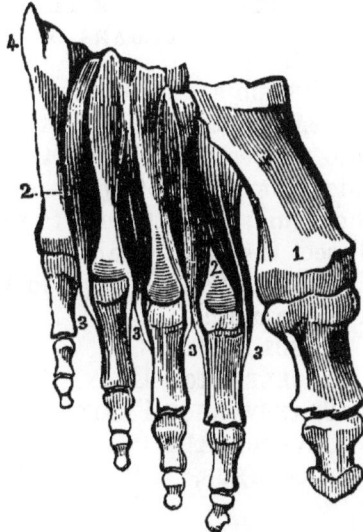

and fleshy, from between the metatarsal bones of the four lesser toes, and are

Inserted, tendinous, into the inside of the base of the first phalanx of each of the three lesser toes.

Use. To move the three lesser toes inward toward the great toe.

INTEROSSEI PEDIS EXTERNI are four in number, larger than the internal interossei, and situated on the back of the foot; they are bicipites, or arise by two slips.

Arise, tendinous and fleshy, between the metatarsal bones of all the toes.

Inserted, the first, abductor indicis pedis, into the inside of the base of the first phalanx of the fore-toe;—the second, adductor indicis pedis, into the outside of the same toe;—the third, adductor medii digiti pedis, into the outside of the middle toe; the fourth, adductor tertii digiti pedis, into the outside of the third toe.

Use. To separate the toes.

CHAPTER XII.

DISSECTION OF THE PERINEUM AND OF THE MALE ORGANS OF GENERATION.

THE subject should be secured as in the operation for stone. A staff passed into the bladder, and some hair into the rectum. Carry two incisions, one upon either side, from the root of the scrotum to the tuberosities of the ischii; from thence to the point of the coccyx, and dissect the flap upward.

The muscles of the perineum consist of five pair, and a single muscle:

ERECTOR PENIS,
ACCELERATOR URINÆ,
TRANSVERSUS PERINEI,   } on each { SPHINCTER ANI,
LEVATOR ANI,   side. {   single muscle.
COCCYGEUS,

The RAPHE, or line running along the skin of the perineum, marks the place where the opposite muscles meet. The appearance of these muscles will vary in different subjects. In those who have died weak and emaciated, the fibres will be pale, and not very evident, while in strong muscular men, who have expired suddenly, they will be very distinct. When the fat and superficial fascia have been carefully cleared away, the first muscle coming into view will be the superficial sphincter ani; and in front of the anus, a shining fascia will be seen stretching across from the ramus of the pubes and ischium of one side to the same points on the other. This is the SUPERFICIAL PERINEAL FASCIA of some, the MIDDLE Perineal fascia of others. All in front of the anus is called the anterior or urethral perineum. This fascia covers the muscles of this part of the perineum, and, instead of continuing back over the superficial sphincter ani, dips down in front of the anus, and becomes connected with the anterior layer of the triangular ligament, or DEEP PERINEAL fascia. On either side of the rectum, and the tuberosities of the ischii, are two large deep fossæ, filled with granulated fat, and having bloodvessels and nerves from the Internal Pudic Artery, vein, and nerves. These are the ISCHIO-RECTAL FOSSÆ. The outer boundary of these cavities is formed by the obturator internus muscles, covered by the obturator fascia; the inner boundary by the levator ani muscles, covered by the fascia of the same name. Clean off the fascia, and you will expose the following muscles:

1. The ERECTOR PENIS—*Arises*, tendinous and fleshy, from the tuberosity of the os ischium; its fleshy fibres proceed upward over the crus of the penis, adhering to the outer and inner edges of the ascending ramus of the os ischium, and of the descending ramus of the os pubis;—but before the two crura meet to form the body of the penis, it ends in a flat tendon, which is lost in the strong tendinous membrane that covers the corpus cavernosum.

Situation. This muscle covers all the surface of the crus penis that is not in contact with bone.

Use. It was formerly supposed to compress the crus penis, and thereby to propel the blood into the forepart of the corpus cavernosum; and to press the penis up-

Fig. 74.

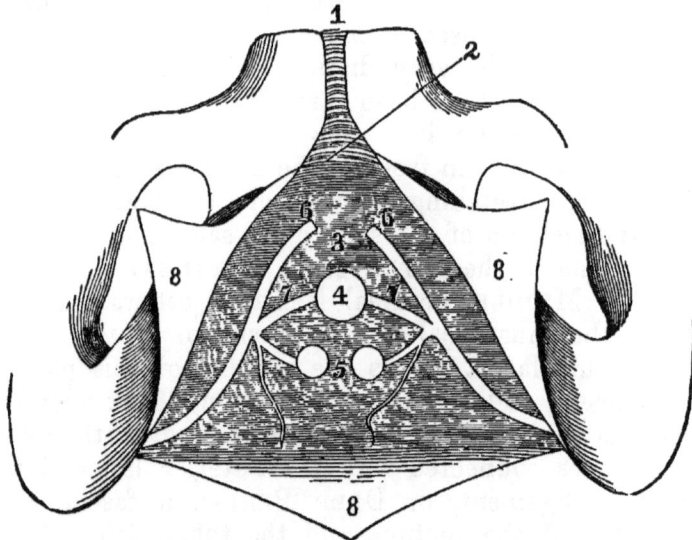

VIEW OF THE DEEP PERINEAL FASCIA.

1. Symphysis Pubis.
2. Sub-pubic Ligament.
3. Triangular Ligament, or Deep Perineal Fascia.
4. Perforation for the Urethra.
5. Two prominences of anterior layer of the Fascia, marking the Position of the included Cowper's Glands.
6. Pudic Arteries.
7. Arteries of the Bulb.
8, 8, 8. The Superficial Perineal Fascia dissected off in three angular Flaps.

ward against the pubis. But its obvious effect must be that of drawing the crus downward to the tuber ischii; which cannot have any influence in contributing to the erect state of the organ.

2. ACCELERATORES URINÆ—*Arise* from a tendinous point in the centre of the perineum, and a tendinous line in the middle of the bulb. The fibres diverge. The inferior ones

Inserted into the ramus of the ischium and pubis.

The middle surround the corpus spongiosum, and the anterior ones extend upon the corpus cavernosum.

Use. To drive the urine and semen forward, by compressing the lower part of the urethra, and to propel the blood toward the corpus spongiosum and the glans penis.

Fig. 75.

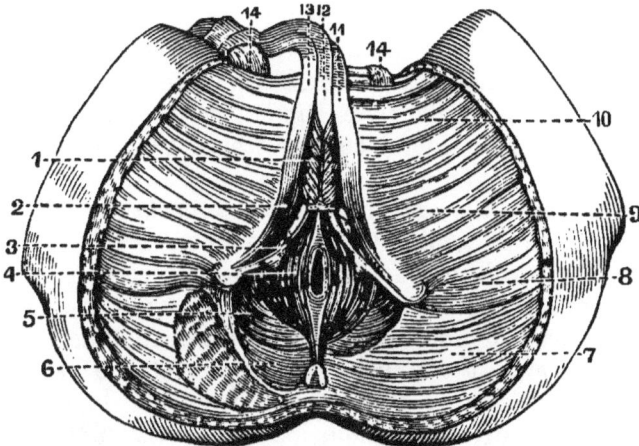

PERINEAL MUSCLES OF THE MALE.

1. Accelerator Urinæ.
2. Erector Penis.
3. Transversus Perinæi.
4. Sphincter Ani.
5. Levator Ani.
6. Coccygeus.
7. Gluteus Maximus.
8. Adductor Magnus.
9. Gracilis.
10. Adductor Longus.
11, 13. Corpora Cavernosa.
12. Urethra.
14, 14. Spermatic Cords.

3. The TRANSVERSUS PERINÆI — *Arises* from the tuber ischii, immediately behind the attachment of the erector penis; thence its fibres run transversely inward.

Inserted into the central point of union where the sphincter ani touches the accelerator urinæ, and where a kind of tendinous projection is formed, common to the five muscles.

Use. To dilate the bulb of the urethra, to prevent the anus from being too much protruded, and to retract it when protruded.

The middle perineal fascia passes behind these muscles

to join the deep or triangular ligament. Accompanying these muscles, is the Transversales Perinei arteries (from the superficial perineal arteries), which, coming from the internal pudic, pass up in the groove between the erectores penis and acceleratores urinæ.

There is sometimes another slip of fibres, the TRANS-VERSUS PERINEI ALTER, which has the same course, and is inserted into the posterior part of the bulb of the urethra.

4. The SPHINCTER ANI EXTERNUS consists of two semicircular planes, which run round the extremity of the rectum, passing nearly as far out as the tuber ischii; the fibres of each side decussate where they meet, and are

Inserted into the extremity of the os coccygis behind; and before, into a tendinous point common to this muscle and to the acceleratores urinæ and transversi perinei. This tendinous point is worthy of remark; it seems to consist in part of an elastic ligamentous substance.

Use. To close the anus, or extremity of the rectum, and to pull down the bulb of the urethra. It is in a state of constant contraction, independently of the will.

5. SPHINCTER ANI INTERNUS—a band of fibres under the superficial sphincter, surrounding the lower end of the rectum.

More deeply seated than the muscles now described, we see some of the fibres of

The LEVATOR ANI.—This muscle *arises* from the inside of the os pubis, at the upper edge of the foramen thyroideum, from the inside of the os ischium, from the tendinous membrane covering the obturator internus and coccygeus muscles; from the semicircular origin its fibres run down like radii toward a centre, and are

Inserted in the two last bones of the os coccygis, and into the extremity of the rectum, passing within the fibres of the sphincter ani, but on the outside of the longitudinal fibres of the gut itself.

Situation. This muscle, with its fellow, very much resembles a funnel, surrounding the extremity of the rectum, the neck of the bladder (which passes through a slit in its fibres), the prostate gland, and part of the vesiculæ seminales.

Use. To draw the rectum upward after the evacuation of the fæces, to assist in shutting it, and to compress the vesiculæ seminales and other viscera of the pelvis.

6. The COCCYGEUS *arises*, tendinous and fleshy, from the spinous process of the os ischium, and covers the inside of the posterior sacro-sciatic ligament; it forms a thin fleshy belly.

Inserted into the extremity of the os sacrum, and into the lateral surface of the coccygis, immediately before the gluteus maximus.

Situation. It is placed betwixt the levator ani and edge of the gluteus maximus.

Use. To support and move the os coccygis forward, and connect it more firmly with the sacrum.

If the muscles be cleared away, the corpus spongiosum will be found to rest upon a dense membrane, which is placed beneath the arch of the pubes, and between the rami of the pubes and ischii. It is the TRIANGULAR LIGAMENT. This ligament consists of two layers, between which is placed the membranous portion of the urethra, the glands of Cowper, and some muscular fasciculi termed the muscles of Wilson and Guthrie. The internal pudic artery, with nerve, is likewise between its lamellæ, and gives off a large branch, the ARTERIA BULBOSI, and another, the ARTERIA CAVERNOSI. The main trunk finally becomes the ARTERIA DORSALIS PENIS.

The rectum must now be separated from the bladder, and pulled downward. This dissection will expose a great part of the levator ani; the neck and body of the bladder; the prostate gland; the vesiculæ seminales; part of the vasa deferentia; part of the ureters; the urethra, its bulb, and corpus spongiosum; the crura penis, and their origin from the ischium; observe

1. The connection of the bladder and rectum, and the cellular substance interposed between them.

2. The prostate gland, *situated* between the bladder

11

and rectum, surrounding the beginning of the urethra in such a manner that one-third of its thickness is situated above the urethra, and two-thirds below it; its shape is somewhat pyriform.

3. The URETHRA.—The curve should be carefully observed. The urethra begins at the neck of the bladder; it is a continuation of that part of the bladder which in the erect posture is lowest. (1) Its beginning is imbedded in the prostate gland. (2) Its membranous part is quite narrow; situated between the prostate gland and bulb of the urethra, and between the layers of the triangular ligament. (3) The urethra then enters the corpus spongiosum.

4. The CORPUS SPONGIOSUM URETHRÆ consists of a plexus of minute veins covered externally by a thin but uniform fibrous sheet; it surrounds the urethra from a short distance from the bladder to its extremity. At its beginning it forms a considerable body of a pyriform shape, termed the *Bulb of the Urethra;* that part of the bulb which is below the urethra is named the pendulous part of the bulb. The corpus spongiosum has on its anterior the glans penis.

5. The GLANDULÆ ANTEPROSTATÆ, or Cowper's Glands, are two small glands of the size of peas between the layers of the triangular ligament.

6. The VESICULÆ SEMINALES are two soft, whitish, knotted bodies, about three or four fingers' breadth in length and one in breadth, and about three times as broad as thick: *situated* between the rectum and lower part of the bladder obliquely, so that their inferior extremities are contiguous, and are affixed to the base of the prostate gland, while their superior extremities are at a distance from each other, extending outward and upward, and terminating just on the inside of the insertion of the ureters in the bladder. They consist of coiled tubes.

7. The two VASA DEFERENTIA are seen running betwixt the vesiculæ seminales, and united to them and to the base of the prostate. Their union forms the DUCTI EJACULATORII. Observe that part of the bladder left

between these tubes, and connected by cellular substance to the rectum, which is pierced when the bladder is punctured from the latter part, no peritoneum intervening.

Fig. 76.

BASE OF THE BLADDER, WITH THE VESICULÆ SEMINALES, URETERS, AND PROSTATE GLAND.

1. Muscular Structure of the Bladder.
2, 2. Ureters.
3, 3. Vasa Deferentia.
4. Vesicula Seminalis.
5. Same of the opposite side, dissected out to show its tubular character.
6. Efferent Duct of the Vesicula Seminalis, which joins the Duct of the Vas Deferens to form at 7 the Ductus Ejaculatorius.
8. Prostate Gland.
9. Urethra.

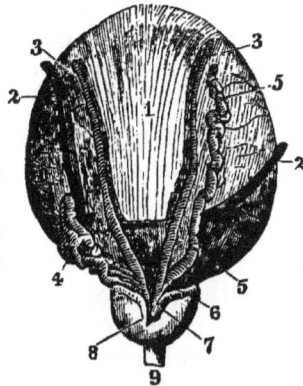

8. The CORPORA CAVERNOSA PENIS arise, on each side, by a process named the CRUS, from the ramus and from the tuber ischii, ascend along the ischium and pubis, and are united immediately before the cartilaginous arch of the pubis. They consist of erectile tissue, covered by a strong, white, fibrous substance, which is very elastic. Internally they are cavernous, and are separated from each other by a septum, which, from being perforated, is named *Septum pectiniforme*.

By the union of the two corpora cavernosa, two grooves are formed: (1) A smaller one above, in which two arteries pass,[1] a large vein or two betwixt them, and some large twigs of nerves. (2) A larger groove below, which receives the urethra.

9. The prepuce is a fold of skin forming a sheath or covering for the glans penis. It makes a duplicature

---

[1] Arteria Dorsalis Penis. It is the termination of the internal pudic artery.

extending along the flat part of the glans from its basis
to the orifice of the urethra, termed FRÆNUM PRÆPUTII.

10. The VESICA URINARIA, or URINARY BLADDER, is
situated within the pelvis, immediately behind the ossa
pubis and before the rectum. It is covered on its upper
and back part by a reflection of peritoneum; in front
and below (where it is contiguous to the rectum) it is
connected by cellular membrane to the surrounding
parts. *Shape*, oval, but flattened before and behind,
and, while in the pelvis, somewhat triangular. *Divided*
into the FUNDUS or bottom, CORPUS or body, and CERVIX
or neck. At the top of the bladder, above the symphysis pubis, may be observed the superior ligament of the
bladder, consisting of the *Urachus*, a ligamentous cord,
which runs up between the peritoneum and linea alba as
far as the navel and two of the ligamentous cords, which
are the remains of the umbilical arteries, and run up
from the sides of the bladder. The bladder is also connected in front by two ligaments formed by the pelvic
fascia, and passing to the viscus from either side of the
symphysis, and on its sides also by the pelvic fascia.

Observe the parts of the bladder not covered by peritoneum, as they are the situations of surgical operations.
These are the whole *anterior surface*, lying against the
pubis, and rising above it, when the bladder is distended,
so that it may be punctured above the pubis; the *sides*,
at the very lowest part of which the cut is made in the
lateral operation of lithotomy, and where the viscus may
be punctured from the perineum: and the inferior surface, resting on the rectum, and allowing us to puncture
from it. Observe also the direction of the axis of the
bladder, in conformity with which all instruments should
be introduced; this is in a line drawn from the navel to
the os coccygis.

11. The entrance of the ureters into the bladder on
the outside of the vesiculæ seminales.

12. The rectum, following the curve of the os sacrum
and os coccygis.

To have a more connected view of the relative situation of these important parts, one side of the pelvis

Fig. 77.

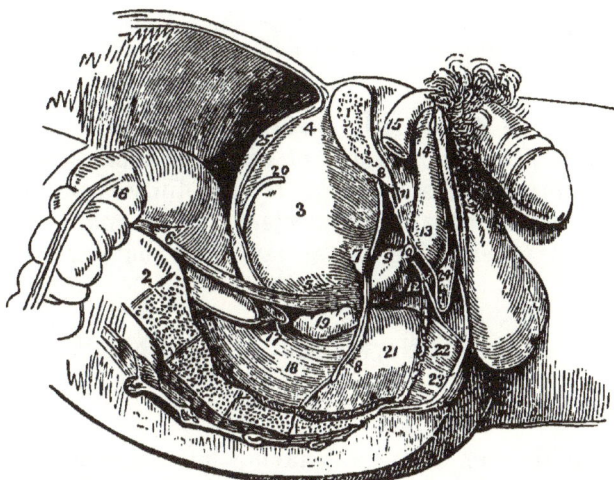

A Side View of the Viscera of the Male Pelvis, *in situ*. The Right Side of the Pelvis has been removed by a Vertical Section made through the Pubis near the Symphysis, and another through the middle of the Sacrum.

1. The Divided Surface of the Pubis.
2. The Divided Surface of the Sacrum.
3. The Body of the Bladder.
4. Its Superior Fundus; from the Apex is seen passing upward the Urachus.
5. The Inferior Fundus of the Bladder.
6. The Ureter.
7. The Neck of the Bladder.
8, 8. The Pelvic Fascia; the Fibres immediately above 7 are given off from the Pelvic Fascia, and represent the Anterior Ligaments of the Bladder.
9. The Prostate Gland.
10. The Membranous Portion of the Urethra, between the two Layers of the Deep Perineal Fascia, or the Triangular Ligament.
11. The Triangular Ligament, or Deep Perineal Fascia formed of two Layers.
12. One of Cowper's Glands between the two Layers of the Triangular Ligament, and beneath the Membranous Portion of the Urethra.
13. The Bulb of the Corpus Spongiosum
14. The Body of the Corpus Spongiosum.
15. The Right Crus Penis.
16. The Upper Part of the Rectum.
17. The Recto-vesical Fold of Peritoneum.
18. The Middle Portion of the Rectum.

19. The Right Vesicula Seminalis.
20. The Vas Deferens.
21. The Rectum covered by the Descending Layer of the Pelvic
    Fascia, just as it is making its bend backward to terminate
    in the Anus.

should now be removed by dividing the symphysis pubis,
and by sawing through the os ilium, or separating it at
its junction with the sacrum. By carefully removing all
the cellular membrane, the student will be enabled more
accurately to examine the situation of the parts above
described.

Of the Vessels and Nerves contained within the Pelvis.

1. *Arteries.*

The A. ILIACA INTERNA, having left the trunk of the
iliaca communis, passes immediately into the pelvis,
where it gives off several large arteries.

1. A. ILEO-LUMBALIS supplies the psoas and iliacus
internus muscles.

2. A. SACRÆ LATERALES, two or three small vessels
which supply the sacrum, cauda equina, and neighboring parts.

3. A. GLUTEA (or *iliaca posterior*), a very large
branch, passes out of the pelvis through the upper part
of the sciatic notch to supply the haunch; but, in its
passage, it gives some branches to the os sacrum, os coccygis, the rectum, and the muscles situated within the
pelvis.

4. A. SCIATICA passes out of the pelvis by the sciatic
notch and below the pyriformis muscle to supply the
hip; in its passage it gives branches to the neighboring
parts.

5. A. PUDICA (*pudenda communis* or *interna*) is the
branch of the internal iliac or ischiatic which is more
immediately destined to supply the parts of generation,
perineum, and lower part of the rectum. It goes out
of the pelvis above the superior sacro-sciatic ligament,
twists round it, and re-enters the pelvis above and before
the inferior sacro-sciatic ligament; it then descends on

the inside of the tuber ischii, ascends on the inner surface of the rami of the ischium and pubis, and, reaching
the root of the penis, divides into two branches. *Superficial and deep.*

Fig. 78.

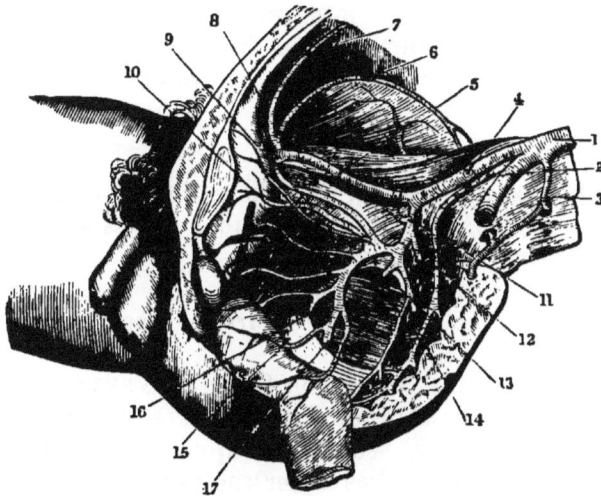

ARTERIES OF THE PELVIS.

1. Termination of the Aorta.
2. Middle Sacral Artery.
3. A Lumbar Artery.
4. Primitive Iliac Artery.
5. External Iliac Artery.
6. Circumflex Iliac Artery.
7. Epigastric Artery.
8. Remains of the Umbilical Artery of the Fœtus converted
   into a Ligament.
9. Obturator Artery.
10. Vesical Artery.
11. Ilio-lumbar Artery.
12 and 13. Lateral Sacral Arteries.
14. Gluteal Artery.
15. Middle Hemorrhoidal Artery.
16. Internal Pudic Artery.
17. Ischiatic Artery.

(1) A. SUPERFICIAL PERINEAL, a branch, which gives
twigs to the bulb of the urethra and neighboring muscles and skin. Noticed in the dissection of the perineum.

(2) ARTERIA BULBOSI, to the bulbous part of the
urethra.

(3) A. CAVERNOSI, to the cavernous bodies.

(4) A. DORSALIS PENIS passes under the arch of the
pubis, runs along the dorsum penis, and is distributed to
the integuments.

While in the pelvis, the pudic gives twigs to the bladder, prostate, and rectum.

2. *Veins.*

The veins attend the arteries and their ramifications; they unite to form the internal iliac vein, except the veins from the rectum, named *hæmorrhoidales*, which ascend along its back part to join the inferior mesenteric vein.

3. *Nerves.*

The nerves met with in this dissection consist of numerous twigs sent off from the lumbar and sacral nerves to supply the parts about the pelvis.

But, in this dissection, we meet with three pair of large nerves, which have their course through the pelvis, and pass to the thigh.

1. Course of the ANTERIOR CRURAL NERVE while in the pelvis. The anterior crural nerve is formed by branches of the first, second, third, and fourth lumbar nerves; at its origin, it lies under the psoas magnus, and, as it descends, passes betwixt the psoas magnus and iliacus internus, till, having passed under Poupart's ligament, it emerges from betwixt those muscles, and appears on the outer side of the femoral artery.

2. Course of the OBTURATOR NERVE within the pelvis. This nerve is formed by branches of the second, third, and fourth lumbar nerves; it lies under the internal border of the psoas magnus, descends into the pelvis, and goes obliquely downward, to accompany the obturator artery through the thyroid hole.

3. Course of the GREAT SCIATIC NERVE within the pelvis. This nerve arises by branches from the fourth and fifth lumbar, and three first sacral nerves, which unite together to form the largest nervous trunk in the body. The nerve passes betwixt the pyriformis and gemini, and thus escapes from the back part of the pelvis by the sciatic notch. Sometimes one of the branches goes through the pyriformis, and joins the sciatic trunk at the back part of the pelvis.

Of the Scrotum.

The scrotum consists externally of a loose, rugose skin, and internally of the Dartos, consisting of unstriped muscular tissue.

On dividing the anterior part of the scrotum, on either side of the raphe, we expose a grayish coat, which is the TUNICA VAGINALIS TESTES. Tunica vaginalis is derived from the peritoneum, carried down in the descent of the testicle; it consists, therefore, like all serous membranes, of two layers, with a cavity within. The outer layer is the tunica vaginalis reflexa; the internal one, the tunica vaginalis propria. Removing this, another very dense coat, the TUNICA ALBUGINEA, within which may be seen the proper glandular structure of the testes, having on its upper edge an appendage termed Epididymis, a little enlarged above and below the GLOBUS MAJOR and MINOR, all of which are convoluted tubes.

2. The Spermatic Cord, connecting the testicle to the abdominal ring. It consists of

a. The spermatic artery, a branch of the aorta; this divides into several branches, which enter the upper edge of the testicle.

b. The spermatic veins, which form a plexus that terminates in the abdomen in a single vein.

c. The spermatic nerves, which come from the sympathetic and lumbar nerves.

d. The vas deferens, or excretory duct of the testicle. This is situated in the back part of the cord, and is distinguished by its firm cartilaginous feel.

e. These parts are all connected by cellular membrane, and by the tunica vaginalis, which is covered by a thin muscle.

f. The cremaster. This arises from the obliquus descendens internus, and is lost on the tunica vaginalis.

It is well now to take out the bladder and penis, and, laying them open by an incision which shall pass through the upper wall of the urethra and bladder, notice the internal appearance. The bladder consists of four coats, the peritoneal one, as has been stated, incomplete; the

11*

others are muscular, cellular, and mucous.   The muscular coat consists of fibres running in different directions. The mucous coat is generally found thrown into folds in the undistended state of the organ.   Some distance be-

Fig. 79.

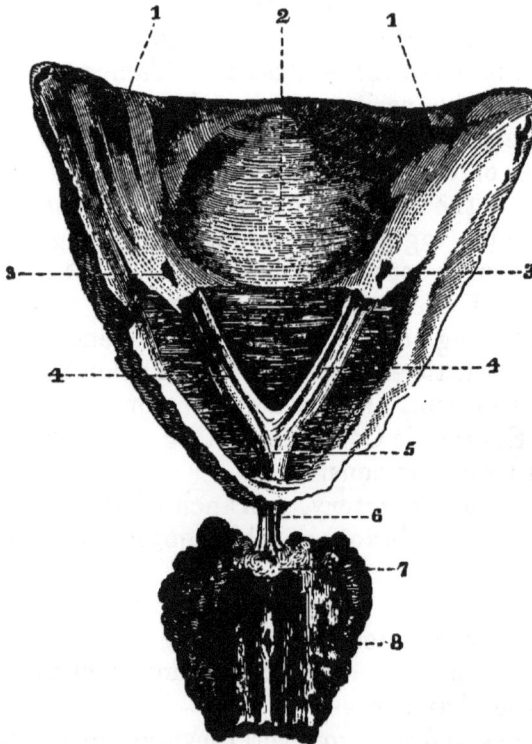

A VIEW OF A PORTION OF THE INSIDE OF THE BLADDER, WITH THE PROSTATE GLAND APPENDED TO IT BY THE ATTACHMENT OF THE COMMON TENDON OF THE MUSCLES OF THE URETERS.

1, 1. Inside of the Bladder.
   2. Lower Fundus.
3, 3. Mouths of the Ureters.
4, 4. Muscles of the Ureters, from which the Mucous Membrane has
       been dissected.
   5. Junction of the Muscles at the apex of the Vesical Triangle.
   6. Tendon of the United Muscles.
   7. Middle Lobe of the Prostate and Point of Insertion, according
       to Sir Charles Bell.
   8. Caput Gallinaginis, the Point of Insertion according to Doctor
       Horner.

hind its neck is a smooth triangular surface, the VESICAL
TRIANGLE. The entrance to the ureters corresponds to
the posterior angles, and the mouth of the urethra to
the apex or anterior angle. The sides of this triangle
are sometimes ridged up ; and, if the mucous membrane
be removed, a few muscular fibres are sometimes seen,
which have been described as the muscles of the ureters.
A pointed projection in the orifice of the urethra is the
uvula vesicæ.

URETHRA.—The first part of the urethra passes
through the prostate gland, which, having in its structure
much unstriped muscular tissue, can compress it. This
portion is called the prostatic portion. In this part we
have a prolonged elevation of its mucous membrane, the

Fig. 80.

THE PROSTATIC, MEMBRANOUS, AND PART OF THE SPONGY PORTION
OF THE URETHRA WITH PART OF THE BLADDER.

1. Internal Surface of the Bladder.
2. Vesical Trigone.
3. Openings of the Ureters.
4. Uvulæ Vesicæ.
5. Urethral or Gallinaginous Crest.
6. Opening of the Sinus Pocularis.
7, 7. Openings of the Ejaculatory Ducts.
8, 8. Openings of the Prostatic Ducts.
    The numbers 7, 7, and 8, 8, are
    placed on the cut surface of the
    Supra-urethral portion of the
    Prostate Gland.
9, 9. Lateral Lobes of the Prostate Gland.
    a. Membranous Portion of the Urethra.
b, b. Cowper's Glands.
c, c. Mouths of the Ducts of the same.
    d. Commencement of the Spongy Por-
    tion of the Urethra.
e, e. Upper Surface of the Bulb.
f, f. Roots of the Cavernous Bodies.
g, g. Corpora Cavernosa.
    h. Spongy Portion of the Urethra.

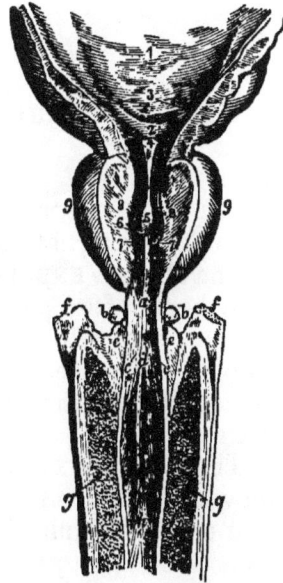

CAPUT GALLINAGINIS ; on its sides are the openings of
the DUCTI EJACULATORII. On each side of the caput
is a considerable depression, the PROSTATIC SINUSES.
The numerous openings upon their floors are those of

the PROSTATIC DUCTS. *Sinus Pocularis:* an opening in front of the Caput Gallinaginis.

MEMBRANOUS PORTION—in advance of the prostatic, and is eight to twelve lines in length.

BULBOUS PORTION—so much of the urethra as traverses the bulb.

SPONGY PORTION—the remaining part of the canal. Just behind its termination is a fossa, the FOSSA NAVICULARIS. Mucous lacunæ are scattered over the surface of the canal.

The GLANDS OF COWPER—two little bodies placed between the layers of the triangular ligament, and opening by two ducts into the anterior part of the bulbous urethra.

CHAPTER XIII.

DISSECTION OF THE ORGANS OF GENERATION IN THE FEMALE.

PREVIOUS to the dissection, it will be proper to examine the external parts.

The MONS VENERIS is a rounded prominence, covered with hairs after puberty. It consists of the common integuments, with an additional quantity of cellular and adipose substance, and lies upon the forepart of the ossa pubis. From the inferior part of the mons veneris arise

The LABIA EXTERNA, called also the labia majora; they are continued downward and forward in the direction of the symphysis pubis and terminate in the perineum anterius: they consist of integuments, cellular substance, and fat, are thicker above than below, and are red and vascular on their inner side. The places where the labia are joined to each other above and below, are termed Commissures; the lower commissure the FOURCHETTE.

The longitudinal cavity, or fissure, situated betwixt the labia, and extending from the mons veneris to the perineum anterius, is sometimes called the SINUS PUDORIS; it is broader above than below, and contains several other parts.

On separating the labia, we see, immediately below the superior commissure,

Fig. 81.

THE VULVA.

1. Mons Veneris.
2. Right Labium.
3. Right Nympha.
4. Clitoris, of which only the Anterior extremity is seen.
5. Vestibule.
6. Orifice of the Urethra.
7. Commencement of the Vagina.
8. Fourchette.
9. Navicularis Fossa.
10. The Anus.
11. Perineum.

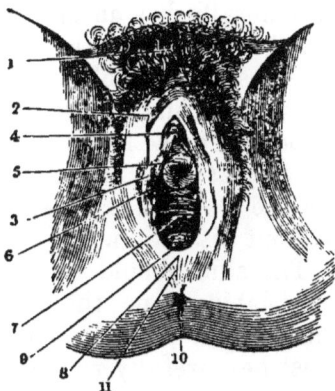

The CLITORIS, a red projecting body, situated below the arch of the pubis, and partly covered by its PREPUCE. The prepuce is a fold of skin, continued from the inner surface of the labia, so as to cover the superior and lateral part of the clitoris. The clitoris resembles the penis of the male, and consists of two cavernous bodies; these, however, cannot be traced in this stage of the dissection. That part of the body which forms an obtuse projection externally, is called the GLANS.

The PERINEUM ANTERIUS is that portion of the soft parts which extend from the inferior commissure of the labia to the anus.

The PERINEUM POSTERIUS is the space betwixt the anus and point of the os coccygis.

LABIA INTERNA, or NYMPHÆ, are two prominent doublings of the integuments, extending from the glans

of the clitoris to the sides of the vagina. Their external side is continued from the inner surface of the labia,
and from the prepuce of the clitoris.

VESTIBULUM—A space bounded above by the clitoridis, and laterally by the nymphæ. At its lower part
we see the orifice of the urethra, above the orifice of the
vagina; it consists of a small rising prominence like a
pea, in the centre of which is a small opening or hole.

The HYMEN, or Circulus Membranosus, is a thin and
extensile membrane, formed by a doubling of the lining
membrane of the vagina, much contracted in .virgins.
It generally has an opening in its upper part.

After the destruction of the hymen, in married women,
we see some irregular projections marking the orifice of
the vagina, and termed CARUNCULÆ MYRTIFORMES.

Behind these is the VAGINA, or canal leading to the
uterus; at the extremity of which may be felt projecting the OS UTERI, or OS TINCÆ, but it cannot be seen
without dissection.

The skin should be now divided on the side of the
right labium, and the dissection should be carried from
the groin to the side of the anus; the cellular membrane
must be carefully removed, in order to expose the following parts.

We find the CLITORIS consisting of two spongy bodies
termed Crura, which unite and form the body. The crus
of each side is a cavernous body, arising from the ramus
and upper part of the tuberosity of the ischium, continued along the ramus of the os pubis, and uniting with
its fellow opposite to the symphysis pubis. The body
formed by the crura does not extend upward, but forms
a curve downward toward the urethra; it is divided internally by the SEPTUM PECTINIFORME, and is attached
to the symphysis pubis by a suspensory ligament; it is
invested by a ligamentous membrane.

The muscles which are met with in this dissection
consist of four pair, and two single muscles.

The ERECTOR CLITORIDIS,
    TRANSVERSUS PERINEI,
    LEVATOR ANI,
    COCCYGEUS,   } on each side.

The SPHINCTER ANI,
SPHINCTER VAGINÆ, } two single muscles.

1. The ERECTOR CLITORIDIS *arises*, fleshy and tendinous, from the tuber ischii, from the inside of the ramus
of the os ischium, and from the ramus of the os pubis;
it passes over the crus of the clitoris, and, becoming
tendinous, is lost upon it.

Use. To draw the clitoris downward and forward,
and, by compressing it, to propel the blood.

Fig. 82.

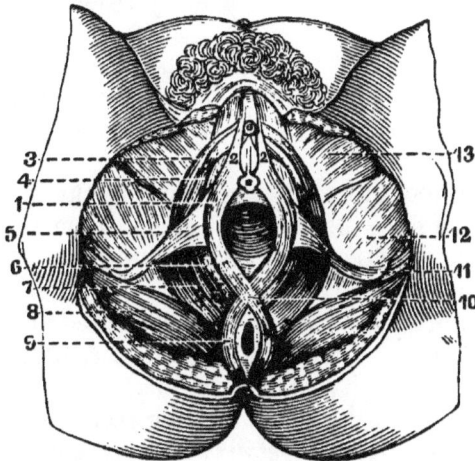

MUSCLES OF THE FEMALE PERINEUM.

1, 2, 6. Sphincter Vaginæ Muscle.
3, 4. Erector Clitoridis   "
5, 11. Transversus Perinæi "
7. Levator Ani   "
8. Gluteus Maximus   "
9. Sphincter Ani   "
10. Junction of the Sphincter Ani and Sphincter Vaginæ Muscles.
12. Adductor Magnus.
13. Gracilis.

Arising from the same point and surrounded by much
cellular membrane, we find

2. The TRANSVERSUS PERINEI.—Its *origin* is the
same as in the male.

It is *inserted* into a ligamentous substance in the perineum anterius, at the point where the sphincter ani
and sphincter vaginæ meet.

This ligamentous or tendinous substance deserves at-

tention.   Here, as in the male, it is the point of union
into which muscles are inserted.

Use. To sustain the perineum.

3. Surrounding the extremity of the vagina, and a
small part of the vestibulum, we find the SPHINCTER
VAGINÆ; it *arises*, anteriorly, from the crura of the clitoris and pubis on each side; it surrounds the orifice of
the vagina, and is

Inserted into the ligamentous point of the perineum.

Use. To contract the mouth of the vagina, and compress the plexus retiformis.

4. The SPHINCTER ANI exactly resembles the same
muscle in the male.

5. The LEVATOR ANI resembles the same muscle of
the male; it surrounds the sides of the vagina in part,
and consequently assists in constricting and supporting it.

6. The COCCYGEUS is longer than in the male.

Under the fibres of the sphincter vaginæ you will find
the PLEXUS RETIFORMIS, or CORPUS SPONGIOSUM VA-
GINÆ, a spongy body, consisting of cellular substance,
interwoven with a number of convoluted bloodvessels.
It arises from the sides of the clitoris, passes on each
side of the extremity of the vagina.

The VAGINA is the canal leading from the vestibulum
to the uterus.   It lies betwixt the rectum and inferior
surface of the urethra and bladder, and is connected to
them by cellular membrane.   It is composed of fibro-
elastic substance, very vascular; its inner surface is rugose, and occupied by mucous glands.   On slitting it up,
we see, at its posterior extremity, the Os Uteri, a rounded
projection, with a transverse fissure.

The UTERUS, or WOMB.—This organ is best seen from
the cavity of the abdomen.   It is situated betwixt the
bladder and rectum, to both of which it is connected by
reflections of peritoneum; it is of the shape of a pear,
and of a firm consistence.   The broad upper part of the
womb is called the Fundus Uteri, the narrower part is
named the neck, or Cervix Uteri, and the intermediate
part its Body.

If the uterus be slit open, its cavity will exhibit the following appearances: Its entrance, the Os EXTERNUM, moderately large.   Then a constricted portion, CERVIX, which expands into a triangular cavity, at the upper angles of which open the Fallopian Tubes.

Fig. 83.

TRANSVERSE SECTION OF THE UTERUS AND PART OF THE VAGINA.

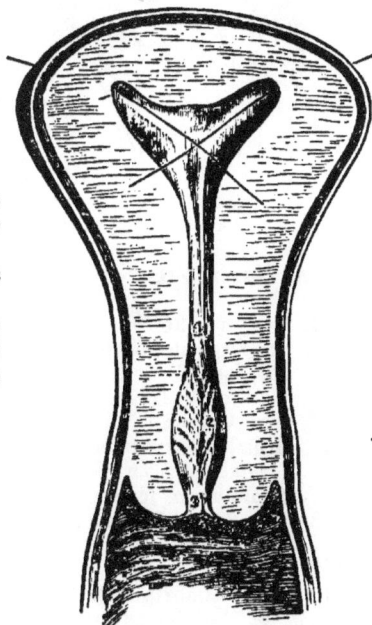

1. Cavity of the Body.
2. Cavity of the Neck, its walls marked by fine oblique ridges.
3. Cervico-vaginal Orifice (Os Uteri).
4. Cervico-uterine Orifice.   The two bristles are introduced through the Orifices of the Fallopian Tubes.

The uterus has four ligaments, two on each side:

The LIGAMENTUM TERES, or Round Ligament.   It is a round long cord, extending from the side of the fundus uteri, and passing through the abdominal ring, to be lost in the groin.

The LIGAMENTUM LATUM, or Broad Ligament, is a broad fold of peritoneum, reflected from the body of the uterus, and connecting it on the sides of the pelvis.   The duplicature of the broad ligament incloses the Fallopian tube, ovary, and round ligament.

The FALLOPIAN TUBES are two.   Each tube is contained in the upper part of the doubling of the broad ligament; it goes out from the fundus of the womb, and is a slender hollow tube. Its outer end is curved downward and backward, and terminates by a broad fringed extremity, termed MORSUS DIABOLI, or the FIMBRIÆ. This broad extremity is connected to the next pair of organs.

Fig. 84.

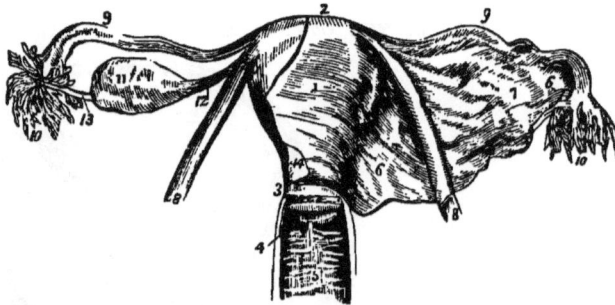

ANTERIOR VIEW OF THE UTERUS AND ITS APPENDAGES.

1. Body of the Uterus.
2. Its Superior Border or Fundus.
3. Its Neck (Cervix).
4. Its Mouth (Os Uteri).
5. The Vagina.
6, 6. Broad Ligament formed by the Peritoneum, which has been removed from the Opposite Side.
7. Prominence formed by the Subjacent Ovary.
8, 8. The Round Ligaments, cut where they enter the Internal Inguinal Ring.
9, 9. Fallopian Tubes.
10, 10. Their Fimbriated Extremities—on the Left side the Extremity of the Tube is turned forward, to show its Mouth or Abdominal Orifice.
11. The Ovary.
12. The Utero-Ovarian or Broad Ligament.
13. One of the Processes of the Fimbriated Extremity of the Tube connected to the Ovary.
14. Cut Edge of the Peritoneum on the Anterior Surface of the Uterus—this Membrane is represented here as descending rather lower upon the organ than is really the case.

The OVARIA are two small oval bodies, white and flat, situated by the side of the uterus, and inclosed in the posterior fold of the broad ligament behind the Fallopian tube; each ovarium is connected to the fundus uteri by a short round ligament, Ligamentum Ovarii.

The BLADDER is situated before the uterus, and is described in the preceding chapter.

The URETHRA is short in females (one inch to one and a half long), and near the bladder is surrounded by a spongy substance.

Fig. 85.

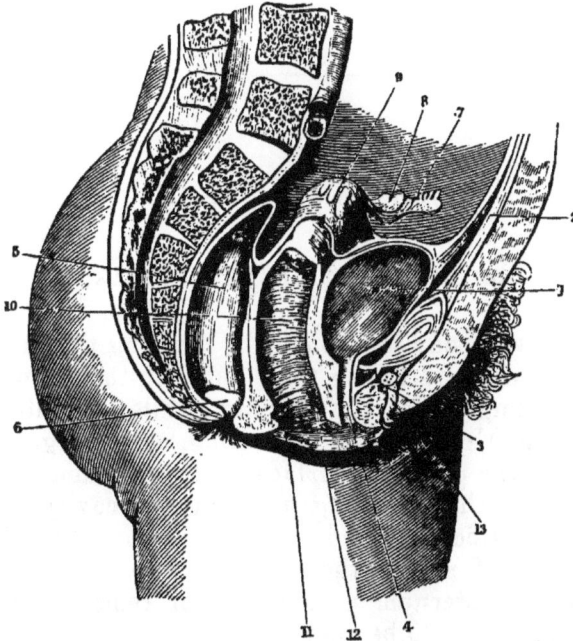

LEFT HALF OF A VERTICAL SECTION OF THE FEMALE PELVIS, WITH THE RECTUM, VAGINA, AND BLADDER LAID OPEN, AND THE UTERUS TURNED TO THE LEFT SIDE.

1. Bladder.
2. Urachus.
3. Anterior Ligament of the Bladder.
4. Urethra.
5. Rectum.
6. Transverse Folds or Pouches of the Rectum.
7. Left Fallopian Tube.
8. Left Ovary.
9. Uterus.
10. Vagina.
11, 12. Anterior and Posterior Vertical Bands or Pillars of the Vagina.
13. Clitoris.

The URETER descends from the kidneys over the psoas muscle; it runs for some space betwixt the bladder and vagina, and at last perforates the bladder near the neck.

The RECTUM lies behind the uterus. (See the preceding chapter.)

To obtain a more satisfactory knowledge of the relative situation of the parts, the left side of the pelvis should be removed as in the male, and the parts examined in that situation.

CHAPTER XIV.

OF PARTS WITHIN THE THORAX.

THE cavity may be exposed by dividing the cartilage from the ribs, and taking these out with the sternum.

On looking under the sternum, while it is lifted up, we see the Mediastinum, separating, as it is gradually torn from the posterior surface of the sternum, into two layers, and thus forming a triangular cavity. This cavity is artificially produced, and is entirely owing to the method of raising the sternum.

When the sternum is laid back or removed, the following parts are to be observed:

The MEDIASTINUM, now collapsed, dividing the thorax into two distinct cavities, of which the right is the largest.

The lungs of each side lying distinct in these cavities.

The pericardium, containing the heart, situated in the middle of the thorax, between the two laminæ of the mediastinum, and protruding into the left side.

The internal surface of the pleura, smooth, colorless, and glistening, lining the ribs, and reflected over the lungs.

1. The PLEURA.—Each side of the thorax has its particular pleura. The pleuræ are like two bladders, situated laterally with respect to each other, by adhering together in the middle of the thorax, and passing ob-

liquely[1] from the posterior surface of the sternum to the dorsal vertebræ, they form the mediastinum. The pleura lines the ribs and the upper surface of the diaphragm, and is reflected over the lung, which is in fact behind it. It forms the LIGAMENTUM LATUM PULMONIS, a reflection of membrane, which connects the inferior edge of the lungs to the spine and diaphragm.

2. The LUNGS.—*Color*, reddish in children, grayish in adults, and bluish in old age. *Shape*, corresponding to that of the thorax, somewhat pyramidal, convex toward the ribs, concave toward the diaphragm, and irregularly flattened next the mediastinum.

Division. (1) The Right Lung is the largest, and is divided into three lobes, two greater ones, and an intermediate lesser lobe.

(2) The Left Lung has two lobes, and also a square notch opposite the apex of the heart. Into the sulci or grooves which form the divisions of the lungs into lobes, the pleura enters; that part of the lung which is affixed to the spine is called its root; it is the part by which the great vessels, nerves, and bronchiæ enter.

3. The PERICARDIUM is a strong, white, and compact membrane, smooth, and lubricated upon the inside, forming a bag for containing the heart, and having its inner lamina reflected over the substance of the heart itself.

4. When you slit open the forepart of the pericardium, you expose the HEART. The right ventricle protrudes; the right auricle, also, is toward you; while the left auricle is retired, and its tip is seen lapping round upon the left ventricle. From under the tip of the left auricle a branch of the coronary vein, and one of the coronary arteries ramify toward the apex of the heart, marking the situation of the SEPTUM CORDIS. The left ventricle will be found firm, fleshy, and resisting, while the right

---

[1] They run obliquely, not being in general attached to the middle of the sternum, but toward its left side, especially at the lower part of the bone, near the diaphragm. Besides the pericardium, the mediastinum contains betwixt its laminæ some adipose membrane and absorbent glands.

ventricle is more loose, and seems partly wrapt round the other.

Fig. 86.

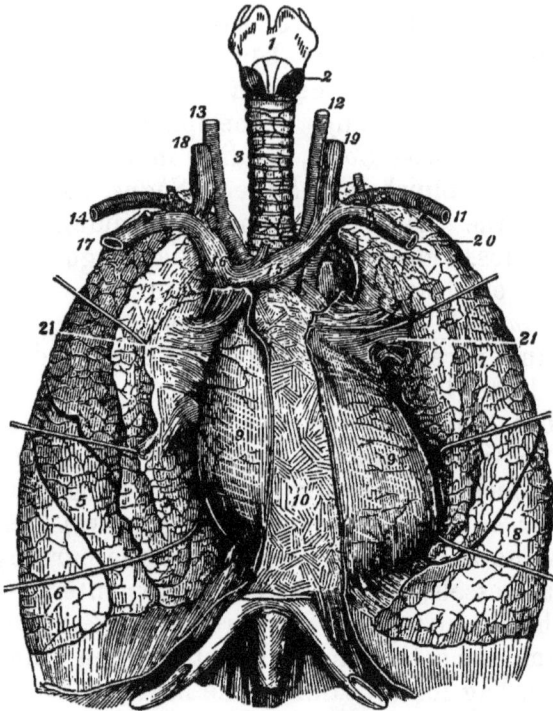

A FRONT VIEW OF THE LARYNX, TRACHEA, AND LUNGS, WITH THE HEART INCLOSED IN THE PERICARDIUM.

1. Thyroid Cartilage.
2. Crico-thyroid Muscle.
3. Trachea
4, 5, 6. Upper, Middle, and Lower Lobes of the Right Lung.
7, 8. Upper and Lower Lobes of the Left Lung.
9, 9. Pericardium investing the Heart.
10. Mediastinum.
11. Left Subclavian Artery.
12. Left Primitive Carotid.
13. Right Primitive Carotid.
14. Left Subclavian Artery.
15. Left Vena Innominata.
16. Right Vena Innominata.
17. Right Subclavian Vein.
18. Right Internal Jugular.
19. Left Internal Jugular.
20. Left Subclavian Vein.
21. Root of the Lungs.
22. Ligamentum Pulmonis.

The heart is situated obliquely in the middle of the breast; its posterior surface is flat, and lies upon the diaphragm; its apex is turned forward and toward the left side, so that, in the living body, it is felt striking between the fifth and sixth ribs, at the point where the cartilages and bony extremities are united. The VENA CAVA SUPERIOR is seen coming down from the upper angle of the pericardium. The INFERIOR CAVA is seen coming up through the diaphragm; but only a very small part of this vein is covered by the pericardium; the two veins enter the right auricle. The RIGHT AURICLE is turned forward, and might be called the anterior; it generally appears black, by the blood shining through its thin coats. The RIGHT VENTRICLE is situated almost directly opposite. The PULMONARY ARTERY arises from the right ventricle; its root is concealed by the right auricle; it ascends on the left side of the aorta; it divides into—1, the right pulmonary artery, which passes under the arch of the aorta, crosses behind it and the vena cava superior to the right lung, and is the longest; and 2, the left pulmonary artery, which passes to the left lung, crossing the descending aorta anteriorly. The PULMONARY VEINS enter the left auricle; two veins come from each lung; the right veins are longest, as they pass behind the vena cava superior. The left auricle is situated on the left side of the right auricle, and somewhat behind it; its tip is seen lapping round upon the LEFT VENTRICLE. This is situated behind, and on the left side of the right ventricle; its substance is stronger and more firm to the touch. The AORTA arises from the back part and right side of the left ventricle; its root is covered by the pulmonary artery. It then ascends betwixt that artery and the vena cava superior. Immediately from the root of the aorta, within the pericardium, the two coronary arteries are sent off to supply the heart itself.

Dissection of the Great Vessels of the Heart.

The VENA CAVA SUPERIOR will be seen descending before the root of the lungs, and on the right side of the aorta. Immediately before perforating the pericardium,

Fig. 87.

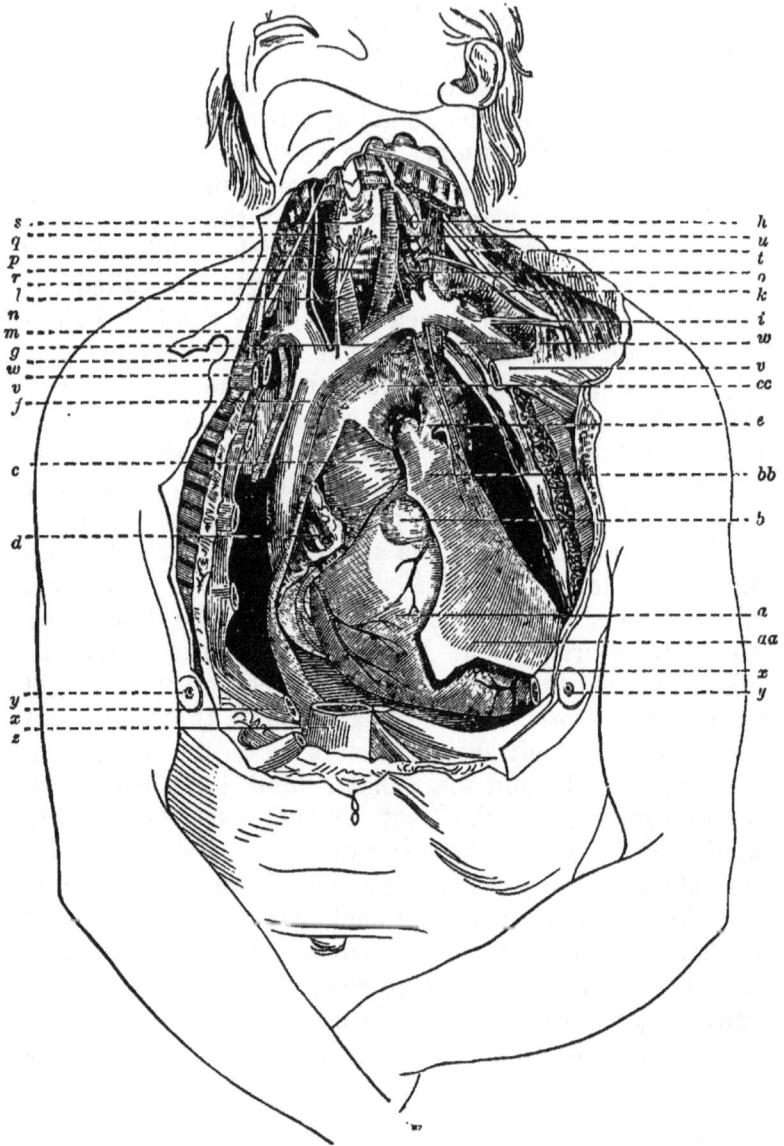

a. Right Ventricle of the Heart.

a, a, and b, b. Pericardium.

b. Pulmonary Artery.

c, c. Arch of Aorta.

d. Right Auricle.

e. Fibrous Remains of the Ductus Arteriosus through which the Pulmonary Artery of the Fœtus communicated with the Aorta.

f. Superior Cava.

g. Left Brachio-cephalic Vein.

h. Left Common Carotid Artery.

k. Lower End of the Left Internal Jugular Vein.

l. Right Jugular Vein.

m. Right Subclavian Vein.

n. Innominata or Brachio-cephalic Artery.

o. Left Subclavian Artery.

p. Right Subclavian Artery crossed by the Pneumogastric Nerve.

q. Right Common Carotid Artery.

r. Trachea.

s. Thyroid Gland.

t. Brachial Plexus of Nerves.

u. Upper End of Left Internal Jugular Vein.

v, v. Clavicles cut across and displaced downward.

x, x. Fifth Ribs cut across.

y, y. Right and Left Breasts.

z. Lower End of Sternum.

it is joined upon its posterior part by the vena azygos, which comes forward from the spine, returning the blood from the intercostal spaces.

Behind the sternum, and just above the arch of the aorta, the superior cava is seen receiving two great branches.

1. A branch coming from the right side, formed by the right subclavian vein, and the right internal jugular.

2. A larger branch coming from the left side (VENA TRANSVERSA or INNOMINATA). It is formed by the left subclavian and left internal jugular, which unite to form a trunk. This trunk crosses before the arteries arising from the arch of the aorta, and then enters the superior vena cava. Into the posterior part of the angle formed by the union of the left subclavian and the left jugular, the thoracic duct empties itself.

On each side the internal jugular vein descends along the neck by the side of the carotid, while the subclavian vein comes from the arm.

The VENA CAVA INFERIOR, immediately after passing through the diaphragm from the abdomen, enters the pericardium.

The AORTA leaves the heart opposite the fourth dorsal vertebra; it crosses over the pulmonary artery, ascends obliquely upward, backward, and to the right side, as high as the second dorsal vertebra. Here it forms an ARCH or incurvation, which passes from the right to the

12

left side, and at the same time obliquely from before
backward. It then comes in contact with the upper
part of the third dorsal vertebra, and descends along
the spine in the posterior mediastinum. This arch of
the aorta is situated behind the first bone of the sternum,
behind and somewhat below the left branch of the vena
cava superior.

From the upper part of the arch come off three large
arteries.

1. The ARTERIA INNOMINATA, or common trunk of
the right carotid and subclavian, ascends above an inch,
and bifurcates into

a. The RIGHT CAROTID, which ascends in the neck by
the side of the trachea.

b. The RIGHT SUBCLAVIAN, which passes outward to
the arm.

2. The LEFT CAROTID.

3. The LEFT SUBCLAVIAN comes off from the extremity of the arch.

The arch of the aorta also gives off some small twigs
which pass to the pleura, the mediastinum, and thymus
gland.

The THYMUS GLAND is a soft glandular body, lying
before the lower part of the trachea and great vessels of
the heart, a little higher than the tops of the two pleuræ.
It is very large in the fœtus, smaller in adults, and nearly
disappears in the aged.

Where the aorta begins to descend, it is connected to
the pulmonary artery by a ligament, which, in the fœtus,
was a large canal, the DUCTUS ARTERIOSUS.

**Dissection of the Posterior Mediastinum,[1] and of the
Nerves and Vessels which have their course through
the Thorax.**

Course of the PHRENIC NERVE through the thorax.—
On each side this nerve is seen entering the thorax betwixt the subclavian artery and subclavian vein. It then

---

[1] By *Posterior Mediastinum* is designed that part of the mediastinum
situated behind the root of the lungs.

proceeds downward and forward before the root of the lungs, and on the outside of the pericardium, betwixt that bag and the pleura. It is lost on the diaphragm. This nerve is accompanied by one artery and two veins. Some twigs pass from the phrenic nerve into the abdomen, to the liver, etc.

Behind the arch of the aorta and great vessels passing from the heart, is seen the TRACHEA. It enters the thorax between the two pleura, and, opposite the third or fourth dorsal vertebra, bifurcates into two parts, BRONCHIÆ, one of which passes toward the right, the other toward the left, to enter the lung of each side.

These bronchiæ divide and subdivide, finally ending in the AIR CELLS. The trachea and larger bronchiæ consist of cartilaginous rings, defective on the posterior third, which is filled up by muscular tissue.

Fig. 88.

TERMINAL VESICLES OF THE LUNG, HANGING TO A BRANCH OF THE
BRONCHIA AS BERRIES HANG TO THEIR STALK.

By folding back the lungs toward the left side of the chest, we expose the pleura reflected from the under surface of the root of the lungs to the spine and ribs. A triangular space is formed betwixt the two pleuræ and the bodies of the dorsal vertebræ. This space or cavity is named the cavity of the posterior mediastinum. It contains many important parts, and must, therefore, be carefully dissected.

But first let us attend to the course of the GREAT SYMPATHETIC NERVE.

CARDIAC PLEXUS.—From the three sympathetic ganglia of the neck come off three nerves, called the SUPE-

RIOR MIDDLE, and INFERIOR CARDIAC NERVES. The superior cardiac receives filaments from the superior laryngeal branch of the par vagum. These nerves, with branches also from the recurrent laryngeal of the par vagum, send filaments about the great bloodvessels at the root of the neck, and afterward form, between the arch of the aorta and the lower part of the trachea, the CARDIAC PLEXUS.

The SYMPATHETIC NERVE, where it enters the thorax, is situated behind the great vessels, close upon the articulation of the first rib with the body of the first dorsal vertebra as it descends along the thorax. It lies upon the heads of the ribs, where they are articulated with the vertebræ. It receives additional branches from all the dorsal intercostal nerves, and in each intercostal space it forms a ganglion. This nerve may be dissected with greater facility when the lungs are removed, and the ribs sawed off near the spine, which will enable the dissector to trace its branches more fully. It lies behind the pleura, but is seen through it. It passes into the abdomen by the side of the spine, running through the fibres of the small muscle of the diaphragm.

Branches of the Sympathetic in the Thorax.

The GREAT SPLANCHNIC NERVE should be attended to. It is formed by twigs, which come off from the sixth, seventh, eighth, ninth, and tenth thoracic ganglia, and penetrates, with the aorta, the diaphragm. LESSER SPLANCHNIC NERVE, formed by filaments from the tenth and eleventh dorsal ganglia, it passes through the crus of the diaphragm, and partly uniting with the great splanchnic, they together terminate in the SEMILUNAR GANGLION, which is formed by a number of smaller ones connected by many filaments, constituting the SOLAR PLEXUS, and placed at the root of the CŒLIAC AXIS.

The SOLAR PLEXUS also receives twigs from the par vagum and the phrenic nerves. The PHRENIC, HEPATIC, SPLENIC, MESENTERIC, SPERMATIC, and RENAL PLEXUSES all emanate from the SOLAR, and are formed by filaments winding about the bloodvessels of these parts.

LUMBAR GANGLIA—four in number, and connecting with the lumbar spinal nerves.

Fig. 89

A.

B.

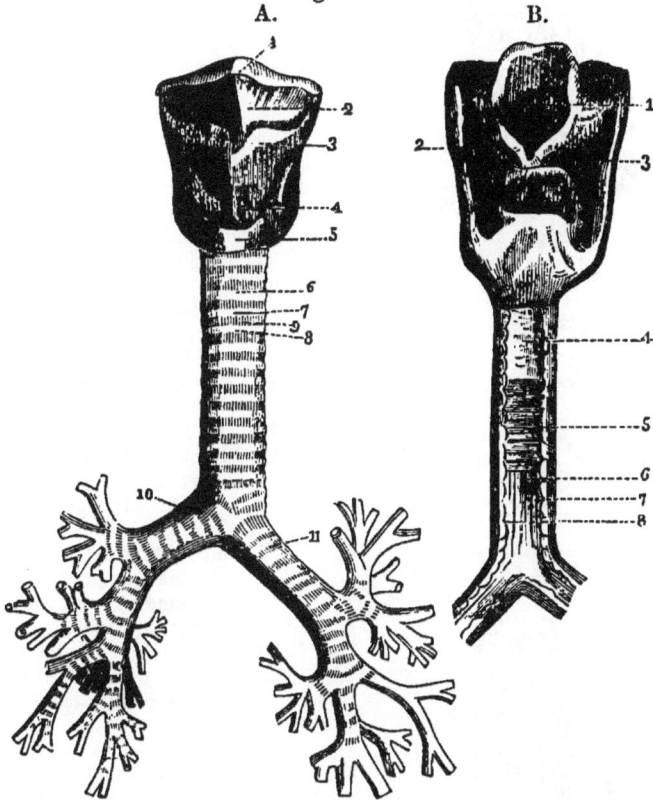

A. FRONT VIEW OF THE LARYNX, TRACHEA, AND BRONCHIAL TUBES.

1. Hyoid Bone.
2. Thyro-hyoid Membrane.
3. Thyroid Cartilage.
4. Crico-thyroid Membrane.
5. Cricoid Cartilage.
6. Trachea.
7, 8. Two Cartilaginous rings.
9. Membrane which separates them.
10. Right Bronchus and its divisions.
11. Left Bronchus.

B. THE LARYNX, TRACHEA, AND COMMENCEMENT OF THE BRONCHIAL TUBES, VIEWED FROM BEHIND.

1. Upper opening of the Larynx.
2, 3. Lateral grooves of the Larynx.
4. Fibrous Membrane of the Trachea, interspersed with small Glands, beneath which is seen
5. The Muscular Fibres; beneath this last are seen
6, 7. Small Fibrous Bands.
8. The Mucous Membrane seen between them.

Fig. 90.

VENA AZYGOS AND THORACIC DUCT.

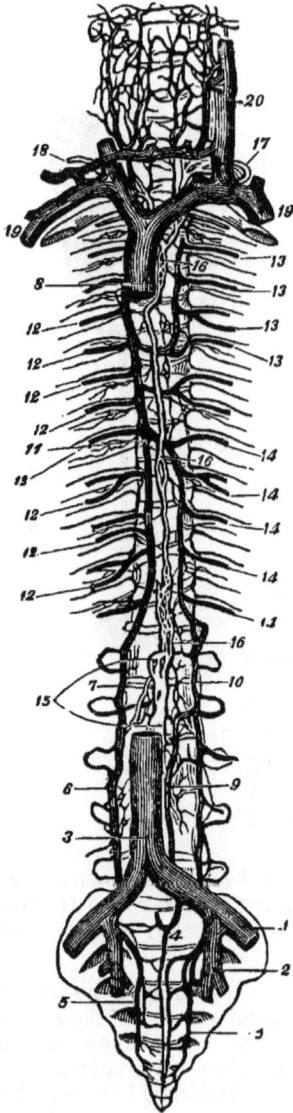

1. External Iliac Vein.
2. Internal Iliac Vein.
3. Ascending Cava.
4. Middle Sacral Vein.
5, 5. Lateral Sacral Veins.
6. Origin of the Greater Vena Azygos in the Lumbar Region and from the Lumbar Veins.
7. Its Trunk.
8. Its Termination in the Descending Cava.
9. Lumbar Veins of the Left Side, forming at
10. The Lesser Vena Azygos, which terminates at
11. In the Greater Azygos.
12, 12, 12. Eight or nine Inferior Intercostal Veins of the Right Side, opening into the Greater Azygos.
13, 13, 13. Superior Intercostal Veins, opening by a common Trunk into the Greater Vena Azygos.
14, 14, 14. Five Inferior Intercostal Veins of the Left Side, joining the Lesser Azygos.
15. Receptaculum Chyli.
16, 16, 16. Thoracic Duct.
17. Its Termination in the Angle formed between the Left Internal Jugular and Left Subclavian Veins.
18. Right Thoracic Duct.
19. Subclavian Vein.
20. Internal Jugular Vein.

HYPOGASTRIC PLEXUS is formed by branches from the lumbar and aortic plexuses, and is distributed to

the pelvic viscera. It is situated at the bifurcation of
the aorta into the iliac arteries, and communicates with
branches from the fourth and fifth sacral nerves.

The sacral ganglia are five in number. The nerves of
the two sides communicate over the coccyx, forming a
ganglion, the GANGLION IMPAR. The cranial ganglia,
six in number, are not described, as the student rarely
pursues such a dissection in the limited time which he
has for his dissecting-room duties.

Toward the middle of the spine you see the VENA
AZYGOS. In dissecting, it is found situated betwixt the
right sympathetic nerve and the aorta; it begins below
from ramifications of the lumbar veins, which pierce the
small muscle of the diaphragm. This vein ascends along
the spine, receiving veins from each of the intercostal
spaces of the right side; and, about the middle of the
back, it receives a considerable trunk, which comes from
under the aorta, VENA AZYGOS MINOR, and returns the
blood from the left side of the thorax. At the fourth
dorsal vertebra, the vena azygos leaves the spine; it
makes a curve forward, and empties its blood into the
back part of the vena cava superior, immediately before
that vein enters the pericardium. The superior intercostal veins on the left side empty into the vena azygos
also.

Descending through the posterior mediastinum will
be also found the AORTA. This great artery, having
formed its arch, comes in contact with the third dorsal
vertebra, and is now called the Descending Aorta, or
Thoracic Aorta. It descends along the bodies of the
dorsal vertebræ, rather on their left side; it lies behind
the œsophagus, and passes betwixt the crura of the
diaphragm into the abdomen.

Branches of the Aorta in the Thorax.

1. The A. INTERCOSTALIS SUPERIOR, on the right
side, is mostly sent off by the subclavian, on the left
side by the aorta.

The Inferior Intercostals are eight or nine in number

on each side of the thorax: they come off separately
from the side or back part of the aorta, and seem to tie
that great artery to the spine. Each intercostal artery
passes immediately into the interval betwixt two ribs,
and there subdivides into

(1) A branch which perforates between the heads of
the ribs to the muscles of the back; this branch also
gives twigs which enter the spinal canal.

(2) The continued trunk of the artery runs forward,
in the interval of the two ribs, giving many branches to
the intercostal muscles. When it reaches the anterior
part of the thorax it is lost in the muscles.

Each intercostal artery is accompanied by one or two
veins, branches of the vena azygos, and by an intercostal or dorsal nerve.

2. A. BRONCHIALES are two, sometimes three, small
twigs of the aorta, one of which passes to the lungs on
each side.

3. Small arteries pass forward from the aorta on the
œsophagus, named A. Œsophageæ; others run to the
pericardium and pleura.

The dissector also finds in the posterior mediastinum
the THORACIC DUCT. He must look for it behind the
œsophagus, betwixt the vena azygos and aorta. It is
collapsed, and appears like cellular membrane condensed, and can only be distinguished when inflated or
injected; it was seen in the abdomen close to the aorta,
and passing into the thorax between the crura of the
diaphragm. It ascends along the posterior mediastinum,
and, about the fourth dorsal vertebra, passes obliquely
to the left side, behind the aorta descendens, and behind
the great arch of the aorta, until it reaches the left
carotid artery. It runs behind this artery and behind
the left internal jugular vein; and, after forming a circular turn or arch, it descends and enters the left subclavian vein at the point where that vein is joined by
the left internal jugular. The absorbents of the right
superior extremity, and of the right side of the head and
thorax, usually form a trunk, which enters the right
subclavian vein.

The ŒSOPHAGUS is also situated betwixt the layers of the posterior mediastinum. It lies immediately before the aorta, but rather toward its left side; it is seen descending from the neck behind the trachea; it passes through an opening in the lesser muscle of the diaphragm, and immediately expands into the stomach.

Behind the trachea and vessels going. to the lungs, and on the forepart of the œsophagus, we meet with a congeries of lymphatic glands. Its muscular fibres are arranged longitudinally and circular.

Course of the Par Vagum, or Eighth Pair of Nerves, in the Thorax.

From the neck, the par vagum passes betwixt the subclavian vein and artery into the thorax; it immediately sends off a large branch, the RECURRENT NERVE, back into the neck. On the right side, this branch twists round under the arteria innominata; on the left side, under the arch of the aorta, it ascends behind the carotid, and lodges itself betwixt the trachea and œsophagus, to both of which it gives branches, and to the muscles of the larynx.

The par vagum, having given off the recurrent, descends by the side of the trachea and behind the root of the lungs. It here sends off numerous filaments to the lungs, which, uniting with twigs from the great sympathetic, form the ANTERIOR and POSTERIOR PULMONARY PLEXUSES. These plexuses lie on the anterior and posterior surfaces of the root of the lungs. Other twigs of the par vagum pass to form the inferior CARDIAC PLEXUS about the pericardium.

The trunk of the eighth pair soon reaches the œsophagus; the left par vagum runs on the forepart of the œsophagus, the right nerve on its back part. Here they split into several branches, which unite again and form a PLEXUS. This plexus is called the ŒSOPHAGEAL. The two nerves continue their course along the œsophagus, and pass with it through the diaphragm, to ramify on the stomach and form the stomachic plexus.

12*

The twelve dorsal or intercostal nerves are also seen in this dissection emerging from the spinal canal, between the bodies of the vertebræ, and supplying the intercostal muscles, etc.

Dissection of the Heart when removed from the Body.

The heart consists of three tunics or coats. 1. An external smooth one, EXOCARDIUM, which is a reflection of the internal lamina of the pericardium. 2. A middle muscular coat. 3. A smooth internal coat, ENDOCARDIUM, which is a continuation of the internal coat of the great veins and arteries. In the right side of the heart we always meet with a considerable quantity of coagulated blood. In the left side there is much less.

Slit open, with the scissors, the two venæ cavæ on their forepart, the inner surface of these veins and of the right auricle will be seen lined by a smooth membrane; and in the auricle the musculi pectinati, or bundles of muscular fibres, will be seen projecting in the auricular appendage. At the point of union between the two cavæ, there is a projection formed by the thickening of the muscular coat, the TUBERCULUM LOWERI. The SEPTUM AURICULARUM is seen separating the right from the left auricle. Observe that it is thin, that in it there is an oval depression, named FOSSA OVALIS. Round this fossa the fibres are thicker, forming the annulus ovalis; this is the remains of the FORAMEN OVALE of the fœtus. The Eustachian Valve is a membrane-like duplicature of the inner coat of the auricle, observed where the vena cava inferior is continued into the auricle, and stretching from that vein toward the opening into the right ventricle. Behind this valve is the orifice of the CORONARY VEIN, with its small valve.

The Foramina Thebesii are minute orifices, some of which are veins, which open into all the cavities of the heart; they are most numerous, however, in the right auricle.

The OSTIUM VENOSUM, or opening of the right auricle

into the right ventricle, is somewhat oval; it has a valve which projects into the right ventricle. .

The RIGHT VENTRICLE may now be opened by an incision, carried from the root of the pulmonary artery down to the apex of the heart. This incision should be made with care, lest the parts on the inside of the ventricle be destroyed by it. It should pass along the right side of the septum ventriculorum, the situation of which is marked out by large branches of the coronary artery and vein. A small opening should first be made, into which one blade of the scissors can be introduced. The incision may be continued through the apex of the heart,

Fig. 91.

A VIEW OF THE INTERIOR OF THE RIGHT AURICLE AND RIGHT VENTRICLE.

1. The Right Ventricle.
2. Tricuspid Valve.
3. Chordæ Tendineæ.
4. Pulmonary Artery.
5. The Aorta.
6. Descending Vena Cava.
7. The Right Auricle.
8. Orifice of the ascending Vena Cava.
9. Vena Cava Ascendens.
10. Valvula Eustachii.
11. Orifice of the Descending Vena Cava.
12. Position of the Tuberculum Loweri.
13. Valvula Thebesii overhanging the orifice of the Coronary Vein.

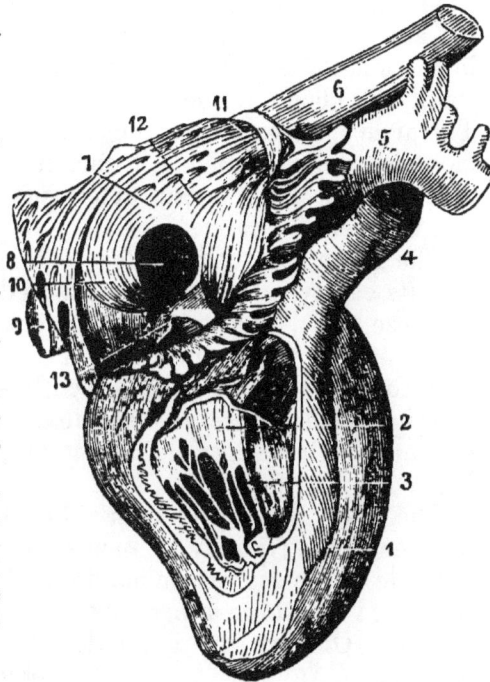

or a flap may be made by another cut, passing from the beginning of the first along the margin of the right auricle. In this ventricle, observe the projecting bundles of muscular fibres, the TRICUSPID VALVES arising from the margin of the ostium venosum, and projecting into the right ventricle. This valve forms a complete circle at its base, but has its edge divided into three parts, which are attached by tendinous filaments, named CHORDÆ TENDINEÆ, to the CARNEÆ COLUMNÆ, or muscular bundles of the ventricle.

The SEPTUM VENTRICULORUM, or partition of the two ventricles, is marked out externally by two veins running from the apex to the basis of the heart.

Slit up the pulmonary artery. Observe how it arises from the back part of the right ventricle, how smooth the inside of the ventricle becomes as it approaches the entrance of the artery, or ostium arteriosum. Observe the three SEMILUNAR or SIGMOID VALVES. Between the valves and wall of the artery are little sinuses, the SINUSES OF VALSALVA. The bases of the valves arise from the artery, their loose edges project into its cavity, and in the middle of the loose edge of each valve is seen a small white body, termed CORPUS SESAMOIDEUM Arantii. The artery is seen bifurcating into the right and left pulmonary arteries, and, just before its bifurcation, sending off to the aorta the ductus arteriosus, which in the adult is a ligament.

The LEFT AURICLE has four pulmonary veins opening into its cavity, which may be exposed by slitting up two of those veins. Observe that its coats are thicker than those of the right auricle. The septum auricularum, with the fossa ovalis, is here seen less distinctly than on the right side. Observe also the ostium venosum, opening into the left ventricle, and giving attachment to the VALVULA MITRALIS.

The LEFT VENTRICLE may be opened in the same manner as the right by an incision carefully made in the left side of the septum or partition of the ventricles, and continued round the upper part of the ventricle under the auricle. Observe the great thickness of the muscular coat; the VALVULA MITRALIS, forming two projections, which are attached by the chordæ tendineæ to the fleshy columns of this ventricle.

Fig. 92.

A VIEW OF THE LEFT VENTRICLE LAID OPEN.

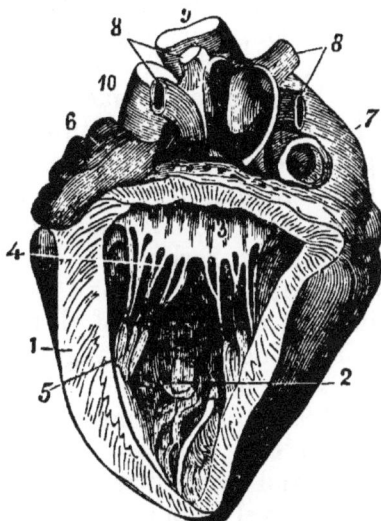

1. Parietes of the Ventricle.
2. Its Cavity.
3. Mitral Valve.
4. Chordæ Tendineæ.
5. Columnæ Carneæ.
6. Right Auricle.
7. Left Auricle.
8, 8. The Four Pulmonary Veins.
9. Aorta.
10. Pulmonary Artery.

Slit up the aorta. It has three semilunar valves, which resemble those of the pulmonary artery. Behind these valves the artery bulges out, as in the pulmonary, forming the SINUSES of the aorta. Above two of the valves lie the orifices of the two coronary arteries, of which the left is the largest.

CHAPTER XV.

DISSECTION OF THE MUSCLES ON THE POSTERIOR PART OF THE TRUNK AND NECK.

AN incision must be made from the occipital protuberance of the occipital along the spine to the top of the sacrum, and the integuments turned off.

In this dissection we meet with twenty-two distinct pairs of muscles, besides a number of small muscles situated between the processes of contiguous vertebræ.

1. The TRAPEZIUS—It *arises*, by a thick round tendon, from the lower part of the protuberance in the middle of the os occipitis behind, and, by a thin tendinous expansion, from the superior transverse ridge of that bone; from the five superior cervico-spinous processes by the ligamentum nuchæ; tendinous, from the two inferior cervical spinous processes, and from the spinous processes of all the vertebræ of the back. The fleshy fibres coming from the neck descend obliquely, while those from the back ascend.

Inserted, fleshy, into the posterior third part of the clavicle; tendinous and fleshy, into the acromion, and into the upper edge of all the spine of the scapula. The fibres slide over a triangular surface at the extremity of the spine of that bone.

Situation. This muscle is quite superficial, and conceals all the muscles situated in the posterior part of the neck and upper part of the back. The LIGAMENTUM NUCHÆ vel COLLI is a ligament which arises from the middle of the occipital bone, runs down on the back part of the neck, adhering to the spinous processes of the cervical vertebræ, and giving origin to the fibres of the trapezius and of other muscles.

Use. To move the scapula in different directions. The superior fibres draw it obliquely upward, the middle transverse ones draw it directly backward, and the inferior fibres move it obliquely downward and backward.

It should be reflected from the spine and head.

2. The LATISSIMUS DORSI—*Arises*, by a broad, thin tendon, from all the spinous processes of the os sacrum and of the lumbar vertebræ; from the spinous processes of the seven inferior dorsal vertebræ from the posterior part of the spine of the os ilium; also from the extremities of the four inferior ribs, by four distinct fleshy digitations, which intermix with those of the obliquus externus abdominis. The inferior fleshy fibres ascend obliquely; the superior run transversely. They pass over the inferior angle of the scapula (from which the muscle often receives a thin fasciculus of fibres) to reach the axilla, where they are all collected and twisted.

Fig. 93,

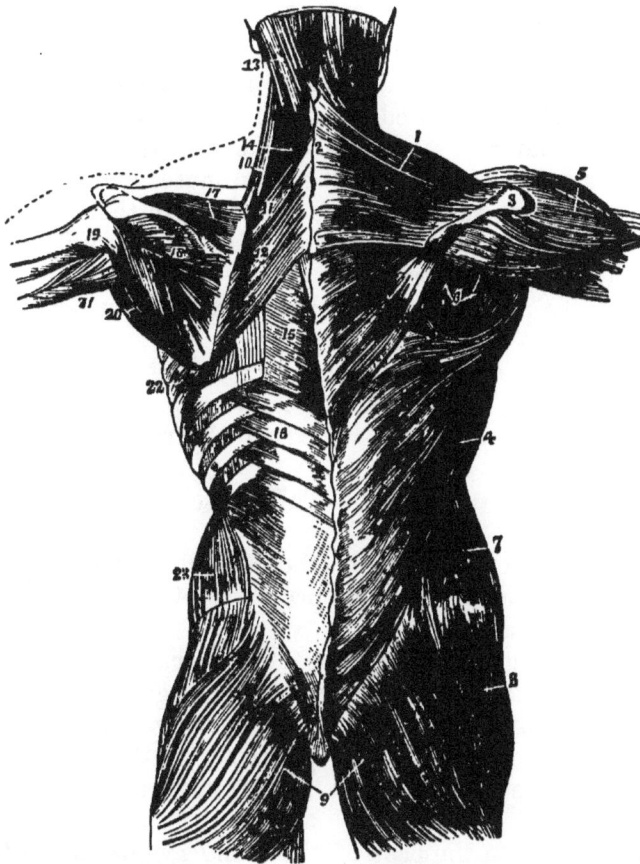

THE FIRST AND SECOND AND PART OF THE THIRD LAYER OF MUS-
CLES OF THE BACK, THE FIRST LAYER BEING SHOWN UPON THE
RIGHT AND THE SECOND ON THE LEFT SIDE.

1. Trapezius Muscle.
2. Tendinous portion of the same, which, with the correspond-ing portion of the opposite Muscle, forms a Tendinous Ellipse on the lower part of the Back of the Neck.
3. Acromion Process and Spine of the Scapula.
4. Latissimus Muscle.
5. Deltoid.
6. Infra-spinatus and Teres Mi-nor Muscles.

7. External Oblique of the Abdomen.
8. Gluteus Medius Muscle.
9. Gluteus Magnus.
10. Levator of the Scapula.
11, 12. Rhomboid Muscles (Small and Large).
13, 14. Splenius Muscle.
15. Aponeurosis covering the Spinal Erector Muscles.
16. Serratus Inferior Posti-cus Muscle.
17. Supra-spinous Muscle.

Inserted, by a strong flat tendon, into the inner edge of the groove in the os humeri, which receives the long tendon of the biceps flexor cubiti.

Situation. Where this muscle arises from the dorsal vertebræ it is concealed by the origin of the trapezius. The remainder of it is placed immediately under the skin, and covers the deeper seated muscles of the loins and back. The tendon of this muscle, with the subjacent tendon of the serratus posticus inferior, assists in forming the fascia lumborum.

Use. To pull the arm backward and downward, and to roll the os humeri.

It should be reflected from the spine, pelvis, and ribs.

Remove the trapezius and latissimus dorsi, and two muscles will be seen passing from the neck to the scapula.

3. The RHOMBOIDEUS.—This muscle is divided into two portions.

(1) Rhomboideus Major (the inferior portion) *arises*, tendinous, from the spinous processes of the four or five superior dorsal vertebræ.

Inserted into all the base of the scapula below its spine, extending as far as its inferior angle.

(2) Rhomboideus Minor (the superior portion) *arises*, tendinous, from the spinous processes of the three inferior vertebræ of the neck, and from the ligamentum nuchæ.

Inserted into the base of the scapula, opposite to the triangular plain surface at the root of the spine.

Situation. This muscle lies beneath the trapezius and latissimus dorsi.

Use. To draw the scapula obliquely upward and directly backward.

The rhomboidei should be reflected from the spine.

4. The LEVATOR SCAPULÆ—*Arises* from the transverse processes of the five superior vertebræ of the neck by five distinct tendinous and fleshy slips, which unite and form a considerable muscle.

Inserted, tendinous and fleshy, into the base of the

scapula, above the root of the spine and under the superior angle.

Situation. This muscle is concealed by the trapezius and sterno-mastoideus; but a small part of its belly may be seen in the space between the edges of these muscles.

Use. To draw the scapula upward and a little forward.

Detaching the rhomboideus from its origin in the spine, you will see another muscle passing from the whole of the basis of the scapula.

5. The SERRATUS MAGNUS—*Arises*, by nine fleshy digitations, from the nine superior ribs. These digitations are seen on the anterior part of the thorax; they pass obliquely backward, and form a strong fleshy muscle.

Fig. 94.

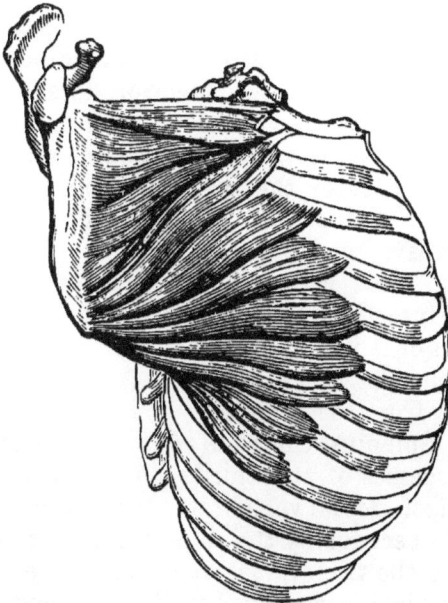

THE SERRATUS MAJOR ANTICUS MUSCLE.

Inserted, principally fleshy, into the whole of the base of the scapula.

Situation. This muscle lies between the scapula and the ribs. The lower digitations, which pass more ante-

riorly than the edge of the latissimus dorsi, are intermixed with the superior digitations of the obliquus externus abdominis.

Use. To move the scapula forward, and, when the
scapula is forcibly raised, to draw the ribs upward.

The removal of the rhomboideus also exposes

6. The SERRATUS SUPERIOR POSTICUS.—This muscle
arises, by a thin, broad tendon, from the spinous processes of the three inferior cervical vertebræ, and of the
two superior dorsal.

Inserted, by distinct fleshy slips, into the second,
third, fourth, and sometimes the fifth ribs, a little beyond their angle.

Situation. This muscle is concealed for the most part
by the rhomboideus and scapula.

Use. To elevate the ribs and dilate the thorax.

Reflect it from the spine.

7. The SPLENIUS is divided into two portions.

(1) The Splenius Capitis—*Arises*, tendinous, from the
spinous processes of the two superior dorsal and five inferior cervical vertebræ.   It forms a flat, broad muscle,
which ascends obliquely, and is *inserted*, tendinous, into
the posterior part of the mastoid process, and into a
small part of the os occipitis, immediately below its
superior transverse ridge.

Situation. This muscle is covered by the trapezius,
and by the insertion of the sterno-cleido-mastoideus, and
a small part of it is seen on the side of the neck betwixt
those two muscles.

Reflect it from the occiput.

(2) The Splenius Colli—*Arises*, tendinous, from the
spinous processes of the third, fourth, fifth, and sometimes the sixth dorsal vertebræ.   It forms a small fleshy
belly, which ascends by the side of the vertebræ, and is
inserted into the transverse processes of the four or five
superior cervical vertebræ, by distinct tendons, which
lie behind similar tendons of the levator scapulæ.

Situation. This muscle is concealed by the serratus
superior posticus and splenius capitis.

Use. To bring the head of the upper vertebræ of the

neck obliquely backward. When both muscles act, they
pull the head directly backward.

Reflect it from the dorsal vertebræ.

8. The SERRATUS POSTICUS INFERIOR—*Arises*, by a
broad, thin tendon, from the spinous processes of the two
or three inferior dorsal vertebræ, and from the three superior lumbar spines by the fascia lumborum.

Inserted, by distinct fleshy slips, into the lower edges
of the four inferior ribs, at a little distance from their
cartilages.

Fig. 95.

1. Splenis Capitis.
2. Complexus Major.
3. Serratus Posticus Superior.

Situation. This is a thin muscle, of considerable
breadth, situated at the lower part of the back, under
the middle of the latissimus dorsi. Its tendon lies under
that of the latissimus dorsi, but, although firmly adher-

ing to it, is distinct, and may be separated by cautious dissection.

Use. To pull the ribs downward and backward.

Reflect it from the spine.

The Fascia Lumborum is now seen. It is a tendinous fascia, arising from the lumbar vertebræ and os sacrum, giving origin to the lower part of the serratus posticus inferior, and to the posterior fibres of the obliquus internus and transversalis abdominis. It is also connected with the tendon of the latissimus dorsi.

On detaching from the spine of this fascia, and the serratus posticus inferior, we expose a thick muscular mass, filling up all the space betwixt the spinous processes of the vertebræ, and the angles of the ribs. This mass consists of three muscles:

(1) Sacro-Lumbalis on the outside.

(2) Longissimus Dorsi in the middle.

(3) Spinalis Dorsi close to the spinous processes.

These three muscles are closely connected together; so that to effect their separation it is necessary to divide some of the fibres.

9 and 10. The SACRO-LUMBALIS and LONGISSIMUS DORSI—*Arise*, by one common origin, tendinous externally, and fleshy internally, from the spinous processes and posterior surface of the os sacrum; from the posterior part of the spine of the os ilium, extending nearly as far forward as the highest part of that bone when the body is erect; from the spinous processes, and from the roots of the transverse processes of all the lumbar vertebræ.

The thick fleshy belly, formed by this extensive origin, ascends, and, opposite to the last rib, divides into the two muscles.

The sacro-lumbalis is *inserted* into all the ribs near their angles, by long and thin tendons. The tendons which pass to the superior ribs are longer, ascend nearly straight, and are situated nearer to the spine than those tendons which pass to the lower ribs. On separating the inner edge of this muscle (*i.e.* the edge next to the spine) from the latissimus dorsi, and turning the belly toward the ribs, we see six or eight small tendinous and fleshy bundles, which pass from the inner side of this muscle, to be inserted into the upper edge of the six or eight inferior ribs. These are called the Musculi Accessorii ad Sacro-Lumbalem.

Fig. 96.

THIRD LAYER OF THE MUSCLES OF THE BACK.

1, 2, 6, 8. Sacro-lumbalis Muscle turned outward to separate it from the Longissimus Dorsi, which lies between it and the spine.
3. Point at which these two muscles are blended in one, the *Sacro-Spinalis.*
4. Complexus Minor.
5. Complexus Major.
7. Transversalis Cervicis.

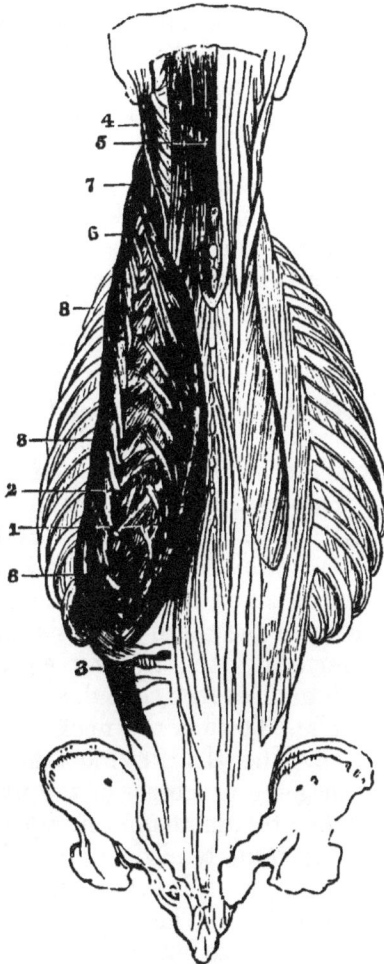

Use. To pull the rib downward, to assist in erecting the trunk of the body, and in turning it to one side.

The longissimus dorsi is *inserted* into all the ribs except the two inferior, betwixt their tubercles and angles, by slips which are tendinous and fleshy, and into the transverse processes of all the dorsal vertebræ by small double tendons.

Use. To extend the vertebræ, and keep the body erect.

11. The SPINALIS DORSI is much smaller than the two last described muscles; below, it cannot be separated from the longissimus dorsi, without dividing some fibres; it lies betwixt that muscle and the spine.

Arises, tendinous, from the spinous processes of the two superior lumbar vertebræ, and of the three inferior dorsal.

Inserted into the spinous processes of the nine upper vertebræ of the back, except the first, by as many distinct tendons.

Use. To extend the vertebræ, and to assist in raising the spine.

12. The CERVICALIS DESCENDENS—*Arises* from the upper edge of the four or five superior ribs by as many distinct tendons, which lie on the inside of the tendinous insertions of the sacro-lumbalis. It forms a small belly, which ascends upward, and is

Inserted, by three distinct tendons, into the fourth, fifth, and sixth cervical vertebræ.

Situation. This muscle is small; it is frequently described as an appendage to the sacro-lumbalis. It arises between the sacro-lumbalis and longissimus dorsi, and is inserted into the transverse processes between the splenius colli and levator scapulæ.

Use. To turn the neck obliquely backward.

13. The TRANSVÉRSALIS COLLI—*Arises* from the transverse processes of the five superior dorsal vertebræ by five tendinous and fleshy slips, and is

Inserted, tendinous, into the transverse processes of the five or six inferior cervical vertebræ.

Situation. The origin of this muscle lies on the inside of the longissimus dorsi, and is sometimes considered as an appendage to it. The insertion is situated between the cervicalis descendens and trachelo-mastoideus.

Use. To turn the neck obliquely backward, and a little to one side.

14. The TRACHELO-MASTOIDEUS lies nearer to the bone than the last described muscle.

Arises from the transverse processes of the three uppermost vertebræ of the back, and of the five inferior of the neck, by as many thin tendons, which unite and form a fleshy belly.

Fig. 97.

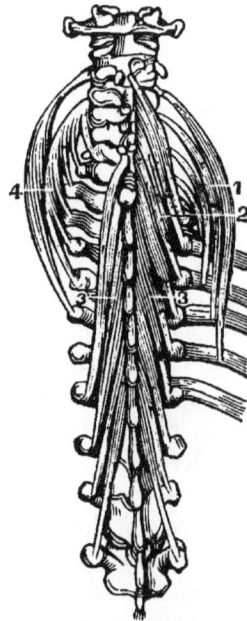

1. Cervicalis Descendens.
2. Semi-spinalis Colli
3. Semi-spinalis Dorsi.
4. Transversalis Colli.

Inserted, tendinous, into the posterior surface of the mastoid process.

Situation. This muscle lies on the outside of the complexus, and on the inside of the transversalis colli; its insertion is concealed by the splenius capitis.

Use. To keep the head and neck erect, and to draw the head backward, and to one side.

15. The COMPLEXUS—*Arises,* by tendinous and fleshy fibres, from the transverse processes of the seven superior

dorsal, and of the four or five inferior cervical vertebræ. It forms a thick, tendinous, and fleshy belly.

Inserted, tendinous and fleshy, into the hollow betwixt the two transverse ridges of the os occipitis, extending from the middle protuberance of that bone, nearly as far as the mastoid process.

Situation. This is a large muscle. Its origin from the cervical vertebræ is nearer to the spine than the trachelo-mastoideus; it is covered by the splenius; but a large portion of it is seen between the splenius and spine, immediately on removing the trapezius.

Use. To draw the head backward, and to one side.

On removing the complexus from the occiput, we find, close to the spine,

16. The SEMI-SPINALIS COLLI.—It *arises,* by distinct tendons, from the transverse processes of the six superior dorsal vertebræ, ascends obliquely close to the spine, and is

Inserted into the spinous processes of all the vertebræ of the neck, except the first and the last.

Situation. This muscle is situated close to the vertebræ at the posterior part of the neck and back.

Use. To extend the neck obliquely backward.

17. SEMI-SPINALIS DORSI—*Arises,* from the transverse processes of the seventh, eighth, and ninth vertebræ of the back, by distinct tendons, which soon grow fleshy.

Inserted, by distinct tendons, into the spinous processes of the five superior dorsal vertebræ, and of the two lower cervical.

Situation. This muscle lies nearer the spine than the lower part of the semi-spinalis colli; its inferior origins lie on the outside of the insertion of the spinalis dorsi.

Use. To extend the spine obliquely backward.

The removal of the complexus brings also in view several small muscles situated at the superior part of the neck, immediately below the occiput.

18. The RECTUS CAPITIS POSTICUS MAJOR—*Arises,* fleshy, from the side of the spinous process of the dentata, or second cervical vertebræ. It ascends obliquely outward, becoming broader, and is

Inserted, tendinous and fleshy, into the inferior transverse ridge of the os occipitis, and into part of the concavity above that ridge.

Situation. This muscle is situated obliquely between the occiput and the second vertebræ of the neck.

Use. To extend and pull the head backward, and to assist in its rotation.

19. The RECTUS CAPITIS POSTICUS MINOR—*Arises*, tendinous and narrow, from an eminence in the middle of the back part of the atlas, or first cervical vertebra. It becomes broader, and is

Inserted, fleshy, into the inferior transverse ridge of the os occipitis, and into the surface betwixt that ridge and the foramen magnum.

Situation. It is partly covered by the rectus capitis posticus major.

Use. To draw the head backward.

Fig. 98.

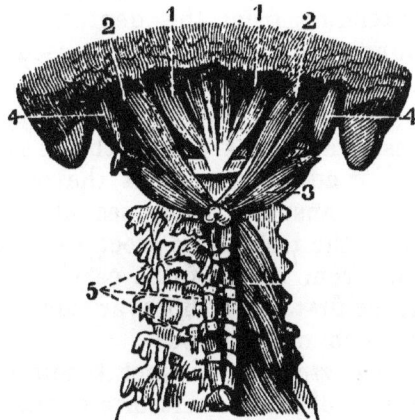

1. Rectus Capitis Posticus Minor.
2. Rectus Capitis Posticus Major.
3. Obliquus Capitis Inferior.
4. Obliquus Capitis Superior.
5. Interspinales.

20. OBLIQUUS CAPITIS SUPERIOR—*Arises*, tendinous, from the upper and posterior part of the transverse process of the first cervical vertebra.

Inserted, tendinous and fleshy, into the inferior transverse ridge of the os occipitis behind the mastoid pro-

13

cess, and into a small part of the surface above and below that ridge.

Situation. This muscle is situated laterally between the occiput and atlas.

Use. To draw the head backward, and to assist in rolling it.

21. OBLIQUUS CAPITIS INFERIOR—*Arises,* tendinous and fleshy, from the side of the spinous process of the dentata or second cervical vertebra. It forms a thick belly, and is

Inserted into the under and back part of the transverse process of the atlas.

Situation. This muscle is obliquely situated between the first two vertebræ of the neck.

Use. To rotate the head, by turning the first vertebra upon the second.

22. The MULTIFIDUS SPINÆ.

On removing the muscles of the spine which have been described, we find situated beneath them the Multifidus Spinæ. It is that mass of muscular flesh which lies close to the spinous and transverse processes of the vertebræ, extending from the dentata to the os sacrum. The bundles of which it is composed seem to pass from the transverse, to be inserted into, the spinous processes.

Arises, tendinous and fleshy, from the spinous processes and back part of the os sacrum, and from the posterior adjoining part of the os ilium; from the oblique and transverse processes of all the lumbar vertebræ; from the transverse processes of all the dorsal vertebræ; and from those of the cervical vertebræ, excepting the three first. The fibres arising from this extensive origin pass obliquely to be

Inserted, by distinct tendons, into the spinous processes of all the vertebræ of the loins and back, and into those of the six inferior vertebræ of the neck. The fibres arising from each vertebra are inserted into the second one above it, and sometimes more.

Use. To extend the back obliquely, or move it to one side. When both muscles act, they extend the vertebræ backward.

The small muscles situated between the processes of the vertebræ are:

1. INTESRPINALES colli, dorsi, et lumborum. These are small bundles of fibres, which fill up the spaces between the spinous processes of the vertebræ. Each of these little muscles arises from the surface of one spinous process, and is inserted into the next spinous process.

In the neck they are large, and appear double, as the spinous processes of the cervical vertebræ are bifurcated. In the back and loins they are indistinct, and are rather small tendons than muscles.

Use. To draw the spinous processes nearer to each other.

2. The INTERTRANSVERSALES colli, dorsi, et lumborum are small muscles which fill up, in a similar manner, the space between the transverse processes of the vertebræ. In the neck they are bifurcated and distinct, in the back they are small and slender, and in the loins they are strong and fleshy.

Use: To draw the transverse processes toward each other.

CHAPTER XVI.

DISSECTION OF THE MUSCLES SITUATED BETWEEN THE RIBS, AND ON THE INNER SURFACE OF THE STERNUM.

THE muscles which fill up the space between the ribs are named Intercostals; they are disposed on each side of the thorax in two layers, and each layer consists of eleven muscles.

1. The INTERCOSTALES EXTERNI—*Arise* from the inferior acute edge of each superior rib, extending from the spine to near the junction of the ribs with their cartilages. The fibres run obliquely forward and downward, and are

Inserted into the upper obtuse edge of each inferior rib, from the spine to near the cartilage of the rib.

Situation. These muscles are seen, on removing the muscles which cover the thorax.

The LEVATORES COSTARUM are twelve small muscles, situated on each side of the dorsal vertebræ. They are portions of the external intercostals. Each of these small muscles *arises* from the transverse process of one of the dorsal vertebræ, and passes downward, to be inserted into the upper side of the rib next below the vertebræ, near its tuberosity.

The first of these muscles passes from the last cervical vertebra, the eleven others from the eleven superior dorsal vertebræ. The three or four inferior Levatores are longer, and run down to the second rib below the transverse process from which they arise.

2. The INTERCOSTALES INTERNI—*Arise* from the inferior acute edge of each superior rib, beginning at the sternum, and extending as far as the angle of the rib. The fibres run obliquely downward and backward, and are

Inserted into the superior obtuse edge of each inferior rib from the sternum to the angle. Portions of the internal intercostals pass over one rib, and are inserted into the next below it.

Thus the intercostal muscles decussate, and are double on the sides of the thorax; but, from the spine to the angles of the ribs, there are only the external intercostals, and, from the cartilages to the sternum, only the internal and some cellular membrane covering them. The whole of the internal intercostals, and the back part of the external, are lined by the pleura.

Use. To elevate the ribs so as to enlarge the cavity of the thorax.

One pair of muscles is situated on the inner surface of the sternum.

The TRIANGULARIS STERNI, or Sterno-Costalis— *Arises*, tendinous and fleshy, from the edge of the whole cartilago-ensiformis, and from the upper edge of the lower half of the middle bone of the sternum. The fibres ascend obliquely upward and outward, and form a flat muscle, which is

Inserted, by three or four triangular fleshy and tendinous terminations, into the cartilages of the third,
fourth, fifth, and sixth ribs.

Situation. This muscle lies on the inside of the ribs
and sternum, and is lined by the pleura.

Use. To depress the cartilages and the bony extremities of the ribs, and consequently to assist in lessening
the cavity of the thorax.

CHAPTER XVII.

DISSECTION OF THE EYE.

The eyes of inferior animals, as the bullock's, sheep's,
or pig's, are generally used for purposes of dissection.[1]

The BALL OF THE EYE—composed of tunics and
humors. The anterior part is covered by a mucous
membrane, the CONJUNCTIVA, which is reflected upon
the eyelids. Clear away all loose structure, and you
expose

The SCLEROTIC COAT—an exceedingly strong fibrous
coat. It is perforated behind by several small openings
for the filaments of the optic nerve, and in front has
connected with it the cornea.

The CORNEA—a transparent membrane, made up of
numerous laminæ. The conjunctiva covers it, though
altered very much in character. Make an incision
through the sclerotica, and, introducing a blowpipe,
force in sufficient air to separate it from the coat beneath; then carefully, with a pair of scissors, divide
around its entire circumference. This accomplished,
make a number of antero-posterior incisions, and turn
back the cut portions, under which the ciliary nerves
will be seen running forward.

---

[1] They should be floated in a saucer of water when being examined.

Tunica Choroidea—a dark coat, depending for its color upon a layer of pigment cells. This coat is very vascular. Where it seems to terminate in front, a white line is seen, the Ciliary Ligament. In front of it is placed the Iris, and, if this be now torn away, a number

Fig. 99.

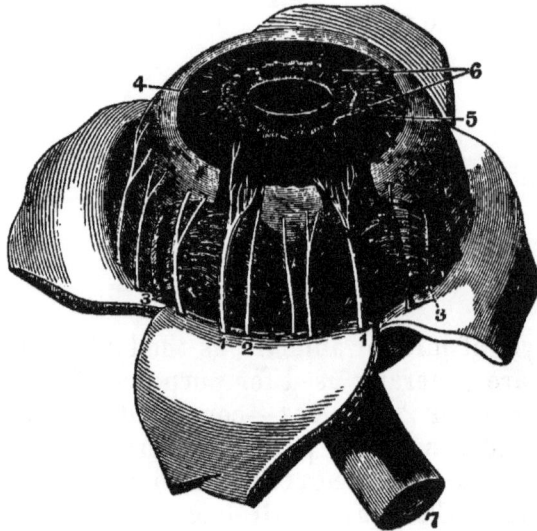

DISSECTION OF THE EYE, IN WHICH THE SCLEROTICA HAS BEEN DISSECTED OFF AND TURNED DOWN IN ORDER TO EXPOSE THE NERVES AND SOME OF THE BLOODVESSELS.

1, 1. Ciliary Nerves entering the Ciliary Ligament and passing forward to the Iris. The Ligament is dissected away in two places to show their course.
2. Smaller Ciliary Nerve.
3. Vasa Vorticosa, or Veins of the Exterior Layer of the Choroid.
4. Ciliary Ligament and Muscle.
5. Converging Fibres of the Greater Circle of the Iris.
6. Looped and knotted form of these Fibres near the Pupil, the knots or enlargements being regarded as Ganglia by Meckel. Within them is seen the *lesser circle*, Sphincter Iridis, with its Converging Fibres.
6. The Optic Nerve.

of vascular fringes will be seen on its front border. These are the ciliary processes.

Fig. 100.

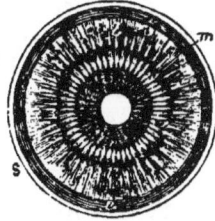

a, b. Corona Ciliaris, or Ciliary Body, the rays
of which are adherent to the Choroid
at *b*, and free at *a*.
s. Sclerotic Coat.
c. Choroid Coat.

The CILIARY LIGAMENT is a line of union between
the iris with the choroidea, and these again with the
sclerotica and cornea.

IRIS—a muscular and vascular body, consisting of
fibres longitudinal and circular. It is colored with pig-
mentum nigrum, and the opening in it is the pupil.

Fig. 101.

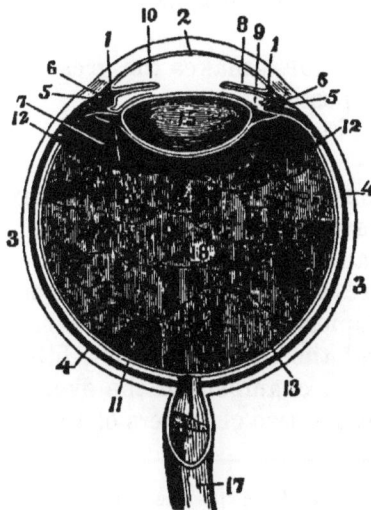

A HORIZONTAL SECTION OF THE EYE.

1, 1. The Cornea, fitted into the Sclerotica.
   2. Its Posterior Lamina, or Cornea Elastica, forming the Anterior Parietes of the Chamber for the Aqueous Humor.
3, 3. Sclerotic Coat.
4, 4. Choroid Coat.

5, 5. Ciliary Ring or Ligament.
6. Its Internal Surface, corresponding to the Ciliary Processes.
7. Ciliary Body, or Corona Ciliaris of the Choroid Coat.
8. The Iris.
9. Posterior Chamber of the Aqueous Humor.
10. Anterior Chamber of the Aqueous Humor.
11. The Retina.
12, 12. Termination of the Retina (according to Cruveilhier and others, *margo dentatus*), at the Posterior Extremities of the Ciliary Processes of the Vitreous Body.
12. The Vitreous Humor.
13. The Hyaloid Tunic, one layer of which passes behind.
14. The other in front of the Crystalline Lens.
15. The Lens.
16 Canal of Petit.
17. Optic Nerve, invested by a Sheath from the Dura Mater.
18. Vitreous Humor, or Corpus Vitreum.

RETINA.—If the choroid be dexterously removed, the retina is exposed, an exceedingly delicate nervous membrane.[1] It is connected in front to the lens by its vascular layer, the ZONULA CILIARIS.

There are three humors or lenses in the interior of the eye.

VITREOUS HUMOR—makes the great bulk of the eye, and is situated posteriorly.

It is inclosed in a delicate capsule, the HYALOID MEMBRANE.

CRYSTALLINE HUMOR is a double convex lens, situated on the front part of the vitreous humor, and inclosed in a membrane called its capsule.

AQUEOUS HUMOR—it has the least consistence of the three, and occupies all the space between the crystalline lens behind and the cornea in front, filling up the anterior and posterior chambers of the eye.   The communication between the two chambers of the eye is the pupil.

---

[1] For its structure and that of the other coats, see works on special anatomy.

CHAPTER XVIII.

LIGAMENTS.

IF the student designs examining the articulations, they should be kept moist after the dissection of the muscles.

The LIGAMENTS are found either in the form of cords, bands, or sacs. The most perfect capsular ligaments are those of the shoulder and hip-joints.

Within the ligaments of movable articulations is a lining of serous membrane. The synovial sac which secretes a tenacious viscid fluid, the SYNOVIA; designed for lubrication of the joints. The student should examine the most important articulations, of which the following may be described:

Articulation of the Lower Jaw.

It is formed between the glenoid cavity of the temporal bone and the head of the inferior maxillary bone.

A CAPSULAR LIGAMENT invests the joint, arising around the glenoid cavity, and inserted about the neck of the inferior maxillary bone. A few additional fibres on the inner and outer side of the capsular ligament have been named the INTERNAL and EXTERNAL LATERAL LIGAMENTS.

STYLO-MAXILLARY LIGAMENT—*Arises* from the styloid process, and is *inserted* on the posterior face of the jaw, close to the angle.

INTERARTICULAR CARTILAGE divides the capsular ligament, and it is seen placed between the condyle and the glenoid cavity. Usually there are two distinct synovial sacs, one above and one below this cartilage.

Fig. 102.

ARTICULATIONS OF THE LOWER JAW.

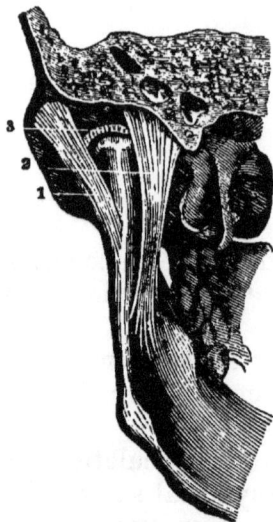

1. External Lateral Ligament.
2. Internal Lateral Ligament.
3. Interarticular Cartilage.

Ligaments of the Spine.

ANTERIOR VERTEBRAL LIGAMENT extends along the front of the spine from the second vertebra to the sacrum.

POSTERIOR VERTEBRAL LIGAMENT extends from the foramen magnum to the sacrum and coccygis, on the posterior part of the bodies of the vertebræ, within the spinal canal.

INTERVERTEBRAL SUBSTANCE. — Fibro-cartilaginous disks placed between all the vertebræ except the first two. They consist of an exterior part, the fibres of which are arranged concentrically, and also oblique, and an interior, consisting of a soft pulpy substance. This material is, in the connected spine, in a state of compression, as it is seen to rise up when the contiguous pieces are removed. The OBLIQUE PROCESSES are connected by capsular ligaments, lined by synovial membranes. The spinous processes have ligaments passing between them. Between the bony bridges of the vertebræ are placed the ligamenta subflava, or yellow elastic ligaments.

Special Articulations of the Spine.

Of the OCCIPUT with the ATLAS.—The articulating processes of each are faced with cartilage, and sur rounded by a capsular ligament.

Fig. 103.

1. Occiput.
2. Posterior Occipito - at- loidean Ligament.
3. Posterior Atloidean Dentate Ligament.
4, 4. Second Pair of Yel- low Ligaments.

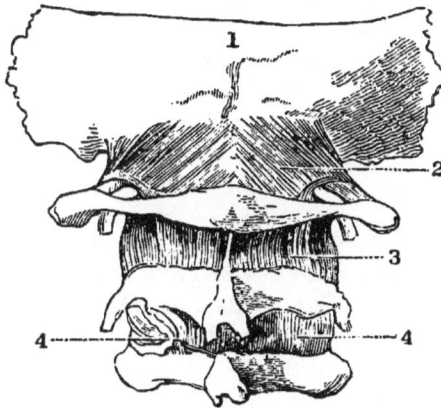

OCCIPITO-ATLOIDEAN LIGAMENT—*Arises* from the margin of the great occipital foramen, and is *inserted* into the upper margin of the Atlas.

Articulation of the Axis with the Occiput.

OCCIPITO-DENTATE LIGAMENT—From the processus dentatus to the front of the great occipital foramen.

TRANSVERSE LIGAMENT stretches across from one side to the other of the first vertebra, just behind the processus dentatus.

MODERATOR LIGAMENTS—Two in number, and extend from the processus dentatus to the inner part of the occipital condyles.

Fig. 104.

THE POSTERIOR ARCH OF THE OCCIPUT AND TWO UPPER VERTEBRÆ.

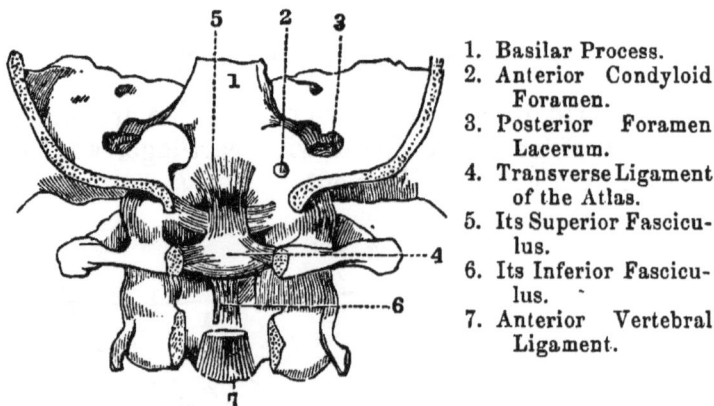

1. Basilar Process.
2. Anterior Condyloid Foramen.
3. Posterior Foramen Lacerum.
4. Transverse Ligament of the Atlas.
5. Its Superior Fasciculus.
6. Its Inferior Fasciculus.
7. Anterior Vertebral Ligament.

Principal Ligaments of the Pelvis.

Those connecting the sacrum to the vertebræ are called SACRO-VERTEBRAL.

The COCCYGEAL LIGAMENTS, ANTERIOR and POSTERIOR, are placed in front and behind the coccyx.

ILIO-LUMBAR LIGAMENT—From the transverse and oblique processes of the last lumbar vertebra, to the posterior part of the crest of the ilium.

SACRO-ILIAC LIGAMENTS—Bands of fibres which surround the sacro-iliac junction.

POSTERIOR SACRO-SCIATIC LIGAMENT — From the posterior inferior spinous process of the ileum, from the sacrum and coccyx, to the inner part of the tuberosity of the ischium, and continued toward the pubis.

ANTERIOR SACRO-SCIATIC LIGAMENT—From the sacrum and coccyx to the spinous process of the ischium.

OBTURATOR LIGAMENT fills up the foramen thyroidcum.

Articulation of the Pubes.

Between the bodies of the pubes, fibro-cartilage, and in front, bands of fibres passing across from one bone to the other.

Fig. 105.

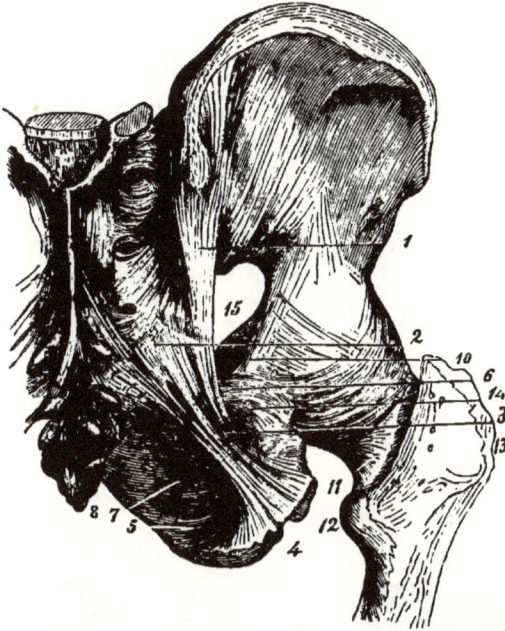

ARTICULATION OF THE PELVIS AND HIP.

1. Posterior Sacro-sciatic Ligament (Vertical Ligament of Bichat), arising from the Sacro-iliac Junction.
2. Also from the Sacrum and Coccyx.
3. Free portion of the Ligament, inserted into the Tuber Ischii at 4 and 5.
6. Lesser or Anterior Sacro-sciatic Ligament.
7. Obturator Ligament.

8. Os Coccygis.
9. Sacral Fasciculus of the Posterior Sacro-Sciatic Ligament.
10, 11. Capsular Ligament of the Hip-joint.
12. Trochanter Minor.
13. Trochanter Major.
14. Lesser Sciatic Notch.
15. Greater Sciatic Notch.
16. Posterior Sacro-iliac Ligament.

SUBPUBIC LIGAMENT placed beneath the arch of the pubes.

Posterior Articulations of the Ribs.

The ribs are connected to the bodies of the vertebræ and intervertebral cartilages by an interarticular ligament, and an anterior one; and to the transverse pro-

cesses by ligaments called COSTO-TRANSVERSE LIGA-
MENTS.

Anterior Articulation of the Ribs.

The ribs have small cavities on their anterior extremities, into which fits the corresponding cartilage, and
strengthened by fibrous bands in front and behind.

The cartilages of the true ribs are let into cavities in
the sternum, and strengthened by radiated fibres in
front.

Sterno-Clavicular Articulation.

The end of the clavicle is placed in a cavity on the
edge of the sternum; a strong cartilage interposed, and
the whole articulation incased in a capsular ligament.

Fig. 106.

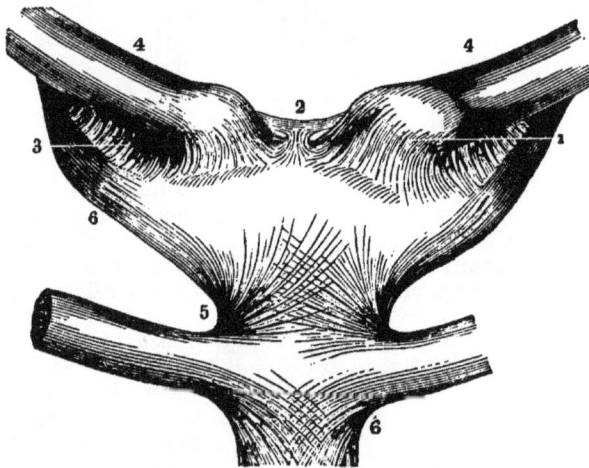

STERNO-CLAVICULAR ARTICULATION.

1. Capsular Ligament.
2. Inter-clavicular Ligament.
3. Costo-clavicular or Rhomboid
   Ligament.
4, 4. Clavicles.
5, 6. Costo-sternal or Chondro-
   sternal Ligaments.

The clavicle is connected also to the first rib or its cartilage, the COSTO-CLAVICULAR or RHOMBOID LIGAMENT.

Scapulo-Clavicular Articulation.

The ACROMIO-CLAVICULAR LIGAMENT.—A capsular ligament investing the acromion process and the acromial end of the clavicle.

CORACO-CLAVICULAR LIGAMENT between the coracoid process of the scapula and the first rib. It presents the appearance of two, which have been named the CONOID and TRAPEZOID LIGAMENTS.

CORACO-ACROMIAL LIGAMENT between the coracoid and acromion processes.

CORACOID LIGAMENT across the coracoid notch.

Scapulo-humeral Articulation.

GLENOID LIGAMENT around the glenoid cavity of the scapula in order to deepen it.

Fig. 107.

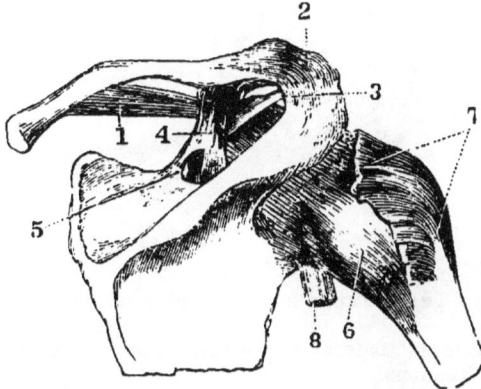

SCAPULO-HUMERAL ARTICULATION.

1. Ligamentum Bicorne.
2. Acromio-clavicular Ligament.
3. Coraco-acromial Ligament
4. Coraco-clavicular Ligament.
5 Coracoid or supra scapular Ligament.
6. Capsular Ligament.
7. Tendons of the Supra-spinatus, Infra-spinatus, and Teres Minor Muscles.
8. Tendon of the Long Head of the Biceps.

CAPSULAR LIGAMENT between the margin of the glenoid cavity and the neck of the humerus.

ACCESSORY LIGAMENT.—A simple thickening of the capsular ligament between the coracoid process and the great tuberosity of the humerus.

The tendon of the biceps muscle passes through the upper part of the cavity on its way through the bicipital groove to the arm.

Elbow-Joint.

CAPSULAR LIGAMENT arising from the margin of the articular surface of the os humeri and inserted into the ulna and coronary ligament of the radius. A thicken-

Fig. 108.

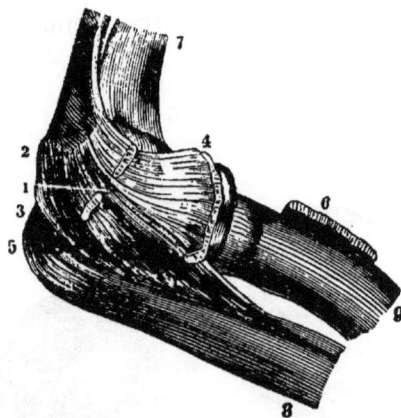

THE HUMERO-CUBITAL ARTICULATION.

1. External Lateral Ligament, blended with the Extensor Tendons.

2, 3, 4, 5. Capsular Ligament.
6. Tendon of the Biceps.
7. Humerus.
8. Ulna.
9. Radius.

ing of its fibres on the inner and outer side of the articulation constitutes the EXTERNAL and INTERNAL LATERAL LIGAMENTS.

CORONARY LIGAMENT OF RADIUS passes from one side of the lesser sigmoid cavity of the ulna to the other. It embraces the neck of the radius.

INTEROSSEOUS LIGAMENT, a fibrous membrane stretched between the radius and ulna.

ROUND LIGAMENT, from the coronoid process to the radius below its tubercle.

Articulation of the Wrist.

This includes the connection of the radius and ulna below; the first row of the carpus with these; and the two rows of the carpus with each other.

Radio-ulnar Articulation.

A triangular cartilage extends from the side of the radius, and is fixed by a pointed process into the root of the styloid process of the ulna. This cartilage separates the lower end of the ulna from the cuneiform bone. The capsule of fibrous tissue which connects these bones above the cartilage is loose-lined with a synovial membrane, and called the SACCIFORM LIGAMENT. The cartilage is often perforated by an opening, and the synovial membrane is continuous above and below.

Radio-carpal Articulations.

Formed between the radius and the first three bones of the first carpal row; it is effected by an ANTERIOR and POSTERIOR RADIO-CARPAL LIGAMENT, and the EXTERNAL and INTERNAL LATERAL LIGAMENT. The lateral ones arising from the styloid processes of the radius and ulna, and inserted into the scaphoid, trapezium, and cuneiform bones.

Articulations of the Bones of the Carpus.

These bones are arranged in two rows, having a common synovial membrane, except the pisiforme, and that between the trapezium and the metacarpal bone of the thumb. The figure will explain. Ligamentous bands run in different directions from one bone to the other.

Fig. 109.

ARTICULATIONS OF THE BONES OF THE CARPUS WITH EACH OTHER, AND WITH THOSE OF THE FOREARM AND METACARPUS.

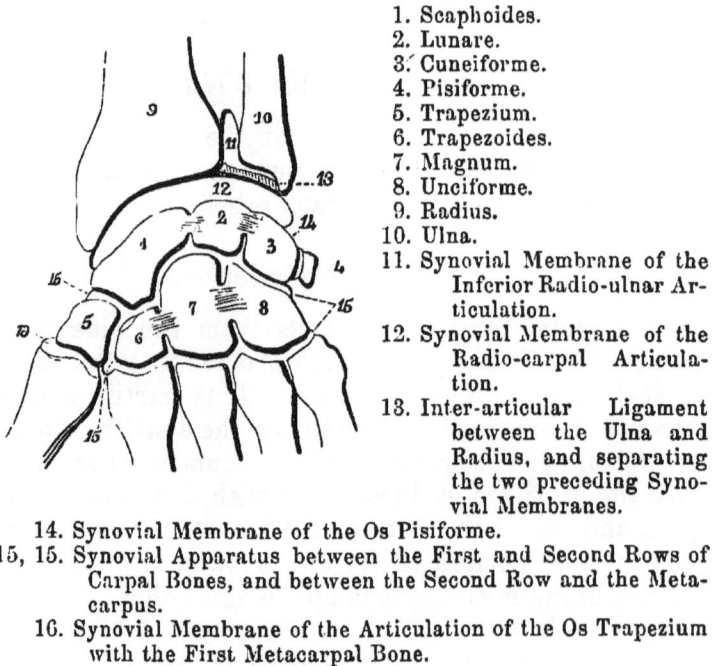

1. Scaphoides.
2. Lunare.
3. Cuneiforme.
4. Pisiforme.
5. Trapezium.
6. Trapezoides.
7. Magnum.
8. Unciforme.
9. Radius.
10. Ulna.
11. Synovial Membrane of the Inferior Radio-ulnar Articulation.
12. Synovial Membrane of the Radio-carpal Articulation.
13. Inter-articular Ligament between the Ulna and Radius, and separating the two preceding Synovial Membranes.
14. Synovial Membrane of the Os Pisiforme.
15, 15. Synovial Apparatus between the First and Second Rows of Carpal Bones, and between the Second Row and the Metacarpus.
16. Synovial Membrane of the Articulation of the Os Trapezium with the First Metacarpal Bone.

The Phalangeal Articulations

may be considered as CAPSULAR LIGAMENTS thickened very much on their sides, forming the LATERAL LIGAMENTS, and strengthened in front and on the back by the sheath for the flexor tendons and the expansion of the extensor tendons:

Articulations of the Hip-Joint.

It occurs between the acetabulum and the head and neck of the femur.

CAPSULAR LIGAMENT—arising from about the acetabulum, and inserted into the neck of the femur, lower
in front than behind.

Fig. 110.

VIEW OF THE CAPSULAR LIGAMENT OF THE HIP-JOINT.

1. The Capsular Ligament
   is separated from the
   Acetabulum, and is
   thrown back to show
   the manner in which
   it invests and con-
   ceals the neck of the
   Femur.
2. Ligamentum Teres.

LIGAMENTUM TERES—a round cord from the pit on
the head of the femur to the sides of the notch of the
acetabulum.

COTYLOID LIGAMENT—surrounds the brim of the
acetabulum as far as the notch.

TRANSVERSE LIGAMENT—subtends the notch of the
acetabulum. The student will observe a loose vascular
pad of fat filling up a little space in the bottom of the
acetabulum. These masses were once considered as
glands.

The Knee-Joint.

An expansion from the tendons of the muscles of the
thigh incloses this articulation in a loose bag of fibrous
tissue.

Fig. 111.

FRONT VIEW OF THE KNEE-JOINT.

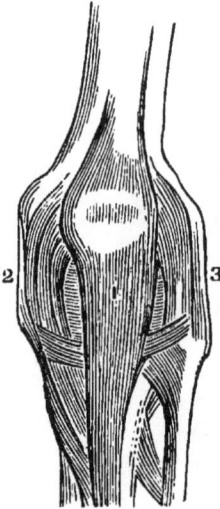

1. Ligamentum Patellæ.
2. Internal Lateral Ligament.
3. External Lateral Ligament.

The Involucrum.

EXTERNAL LATERAL LIGAMENT—from the external condyle to the head of the fibula.

INTERNAL LATERAL LIGAMENT—from the internal condyle of the femur, some distance along the head and upper part of the shaft of the tibia.

In front of the articulation is placed the patella, connected by the TENDO PATELLÆ to the tubercle of the tibia.

LIGAMENT OF WINSLOW—a derivation from the tendon of the semimembranosus muscle, passing to the back part of the capsular ligament and giving it increased strength. If the patella be turned off the front of the joint, a mass of fat is seen filling up the space between the condyles of the femur and the head of the tibia. A ridge of synovial membrane on each side of this forms the ALAR LIGAMENTS. From their junction above starts another duplicature of synovial membrane back to the crucial ligaments. This is the LIGAMENTUM MUCOSUM.

Fig. 112.

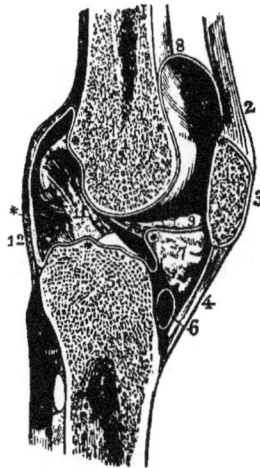

A LONGITUDINAL SECTION OF THE LEFT KNEE-JOINT, SHOWING THE
REFLECTION OF ITS SYNOVIAL MEMBRANE.

1. The Cancellated Structure of the lower part of the Femur.
2. The Tendon of the Extensor Muscles of the Leg.
3. The Patella.
4. Ligament of the Patella.
5. The Cancellated Structure of the Head of the Tibia.
6. A Bursa situated between the Ligament of the Patella and the
   Head of the Tibia.
7. The Mass of Fat projecting into the Cavity of the Joint below
   the Patella. ** The Synovial Membrane.
8. The Pouch of the Synovial Membrane, which ascends between
   the Tendon of the Extensor Muscles of the Leg and the Front
   of the Lower Extremity of the Femur.
9. One of the Alar Ligaments. The other has been removed with
   the opposite ˈsection.
10. The Ligamentum Mucosum left entire—the Section being made
    to its inner side.
11. The Anterior or External Crucial Ligament.
12. The Posterior Ligament. The scheme of the Synovial Mem-
    brane, which is here presented to the student, is divested of
    all unnecessary complications. It may be traced from the
    Sacculus (at 8) along the inner surface of the patella; then
    over the adipose mass (7), from which it throws off the Mu-
    cous Ligament (10); then over the head of the Tibia, form-
    ing a sheath to the Crucial Ligaments; then upward along
    the Posterior Ligament and Condyles of the Femur to the
    Sacculus, where its examination commenced.

CRUCIAL LIGAMENTS, ANTERIOR and POSTERIOR—
the first from the inner face of the external condyle,
and inserted in front of the spinous process of the tibia;
the last from the inner face of the internal condyle of
the femur, and inserted behind the same process of the
tibia.

SEMILUNAR CARTILAGES—two in number, between
the femur and tibia. Their posterior and anterior ends
are fixed behind and in front of the spinous process of
the tibia. Notice the extent of the synovial membrane
above the patella.

Peroneo-tibial Articulations.

The upper end of the fibula is connected to the tibia
by a capsular ligament thick in front, and behind forming the ANTERIOR and POSTERIOR PERONEO-TIBIAL
LIGAMENTS. At the lower end in the same manner, but
called there the ANTERIOR and POSTERIOR INFERIOR
PERONEO-TIBIAL LIGAMENTS.

TRANSVERSE LIGAMENT—extends between the two
malleoli on the posterior aspect of the joint.

INTEROSSEOUS LIGAMENT—fills up the space between
the shafts of the two bones.

The Ankle-Joint.

Constituted by the tibia, fibula, and astragalus. On
the front and behind the joint there exists only the appearance of a capsular ligament, but on the sides we
have the LATERAL LIGAMENTS.

The EXTERNAL LATERAL LIGAMENT consists of three
fasciculi, arising from the external malleolus, and is inserted into the astragalus and the os calcis.

INTERNAL LATERAL LIGAMENT—*Arises* from the internal malleolus, and is inserted into the lesser apophysis of the os calcis. It spreads out toward its insertion,
hence sometimes called the DELTOID LIGAMENT.

ARTICULATIONS. 295

Fig. 113.

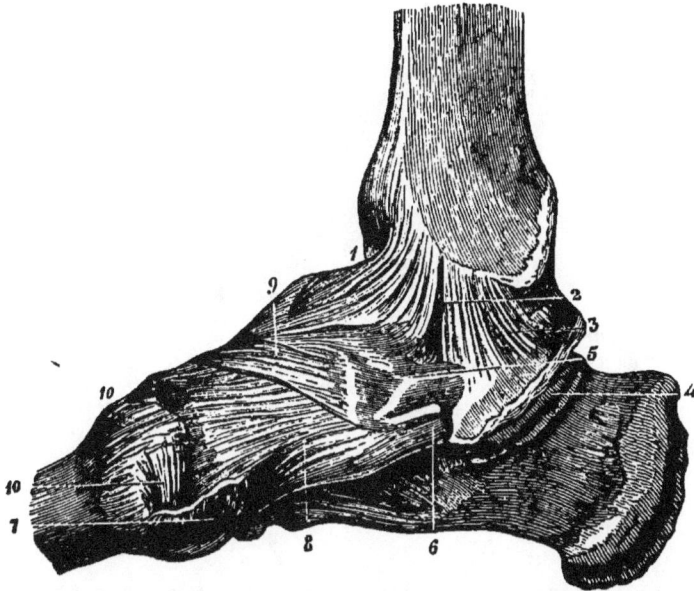

INTERNAL LIGAMENTS OF THE ANKLE AND FOOT.

1. Anterior Fasciculus of the *Deltoid Ligament*.
2. Middle Fasciculus.
3. Posterior Fasciculus.
4. Groove for the Flexor Digitorum Communis.
5. Internal Calcaneo-scaphoid Ligament.
6. Tendon of the Tibialis Posticus.
7. Tendon of the Tibialis Anticus.
8. Ligament connecting the Os Scaphoides with the first Cuneiform Bone.
9. Ligament connecting the Scaphoides with the Cuneiform Medium.
10. Ligaments connecting the first Metatarsal with the first Cuneiform Bone.

Articulations of the Os Calcis and Astragalus.

A very powerful interosseous ligament forms the principal one.

SCAPHOID AND ASTRAGALUS—united by a capsular ligament.

OS CALCIS AND CUBOIDES—united by the superior and inferior calcaneo-cuboid ligaments passing between the two bones.

Fig. 114.

THE EXTERNAL LATERAL LIGAMENTS OF THE ANKLE AND FOOT.

1. Anterior Ligament of the Lower Tibio-fibular Articulation.
2. External Lateral Ligament, sometimes called Peroneo-calcaneum.
3. Anterior Fasciculus of the same, or Peroneo-astragalian Ligament.
4. External Calcaneo-astragalian Ligament.
5. Interosseous Ligament.
6. Lower Calcaneo-cuboid Ligament.
7. Ligament (Ligamentum Dorsale Obliquum), uniting the Fifth Metatarsal Bone with the Os Cuboides.
8. Dorsal Ligament of the Fourth Metatarsal Bone. The dorsal surface of the foot is covered by smaller ligaments that connect the tarsal and metarsal bones, and these again with each other.

The OS CALCIS AND SCAPHOID are very firmly united by the internal and external CALCANEO-SCAPHOID LIGAMENTS—the former from the lesser apophysis of the os calcis to the inner surface of the scaphoid; the latter from the greater apophysis of the os calcis to the outer end of the scaphoid. Other ligaments, both on the upper and lower surface of the foot, connect the different pieces together.

INDEX.

Muscles—*continued.*
    flexor brevis pedis, 216, 217.
        carpi radialis, 112.
            ulnaris, 113.
        digitorum brevis pedis, 191.
            longus pedis, 209.
        longus pollicis manus, 117.
            pedis, 209.
        ossis metacarpi pollicis, 126.
        parvus minimi digiti, 129.
        pollicis longus manus, 117.
        profundus digitorum, 116.
        sublimis digitorum, 114.
    gastrocnemius, 206.
    gemellus, 197.
    genio-hyoglossus, 63.
        hyoideus, 63.
    gluteus maximus, 195.
        medius, 195.
        minimus, 196.
    gracilis, 180.
    helicis major, 13.
        minor, 13.
    hyo-glossus, 64.
    iliacus, 166.
    indicator, 124.
    infra-spinatus, 104.
    intercostales, 275–6.
    interossei manus, 129.
        pedis, 219.
    inter-spinales, 274.
    inter-transversales, 275.
    latissimus dorsi, 262.
    levator anguli oris, 16.
        ani, 224, 238.
        costarum, 276.
        labii superius, 15.
            inferius, 20.
        palati, 81.
        palpebræ, 88.
        scapulæ, 264.
    lineæ longitudinales Lancisii, 37.
        transversa, 37.
    lingualis, 65, 82.
    linguæ superficialis, 82.
        transversalis, 82.

Muscles—*continued.*
        linguæ verticalis, 82.
    longissimus dorsi, 268.
    longus colli, 76.
    lumbricales, 125, 217.
    masseter, 21.
    multifidus spinæ, 274.
    mylo-hyoideus, 62.
    obliquus externus abdominis, 141.
        capitis superior, 273.
            inferior, 274.
        oculi, 89.
        inferior oculi, 90.
        internus abdominis, 144.
    obturator externus, 199.
        internus, 197.
    occipito-frontalis, 11.
    omo-hyoideus, 59.
    orbicularis oris, 19.
        ostium venosum, 258.
        palpebrarum, 14.
    palato-glossus, 80.
        pharyngeus, 80.
    palmaris brevis, 125.
        longus, 112.
    pectinalis, 181.
    pectoralis major, 94.
        minor, 95.
    peroneus brevis, 191.
        longus, 189.
        tertius, 189.
    plantaris, 208.
    platysma myoides, 55.
    popliteus, 208.
    pronator quadratus, 117.
        radii teres, 111.
    psoas magnus, 165.
        parvus, 165.
    pterygoideus externus, 22.
        internus, 23.
    pyramidalis, 150.
        nasi, 11.
    pyriformis, 196.
    quadratus femoris, 198.
        lumborum, 167.
    rectus abdominis, 149.
        capitis anticus major, 77, 272.
            minor, 78, 273.
        lateralis, 78.
        posticus minor, 273.
            major, 272.

THE END.

PUBLICATIONS OF J. B. LIPPINCOTT & CO., Phila.

Will be sent by Mail, post-paid, on receipt of price.

NEW AMERICA.

By WILLIAM HEPWORTH DIXON, Editor of "The Athenæum," and author of "The Holy Land," "William Penn," etc. With Illustrations from Original Photographs. Third Edition. Complete in one volume, Crown Octavo. Printed on tinted paper. Extra Cloth. Price $2.75.

In these graphic volumes Mr. Dixon sketches American men and women, sharply, vigorously, and truthfully, under every aspect. The smart Yankee, the grave politician, the senate and the stage, the pulpit and the prairie, loafers and philanthropists, crowded streets and the howling wilderness, the saloon and the boudoir, with women everywhere at full length—all passes on before us in some of the most vivid and brilliant pages ever written.—*Dublin University Magazine.*

ELEMENTS OF ART CRITICISM.

A Text-book for Schools and Colleges, and a Hand-book for Amateurs and Artists. By G. W. SAMSON, D.D., President of Columbian College, Washington, D. C. Second Edition. Crown 8vo. Cloth. Price $3.50.

This work comprises a Treatise on the Principles of Man's Nature as addressed by Art, together with a Historic survey of the Methods of Art Execution in the departments of Drawing, Sculpture, Architecture, Painting, Landscape Gardening, and the Decorative Arts. The *Round Table* says: "The work is incontestably one of great as well as unique value."

HISTORY OF THE U. S. SANITARY COMMISSION.

Being the General Report of its Work on the War of the Rebellion. By CHARLES J. STILLÉ, Professor in the University of Pennsylvania. One vol. 8vo. Cloth, beveled boards. Price $3.50.

TERRA MARIÆ; or, Threads of Maryland Colonial History.

By EDWARD D. NEILL, one of the Secretaries of the President of the United States. 12mo. Extra Cloth. Price $2.00.

COMING WONDERS, expected between 1867 and 1875.

By the Rev. M. BAXTER, author of "The Coming Battle." One vol. 12mo. Cloth. Price $1.00.

A TUTOR'S COUNSEL TO HIS PUPILS.

En Avant, Messieurs! Letters and Essays. By the Rev. G. H. D. Mathias, M.A. Second Edition. Small 12mo. Extra Cloth. Price $1.50.

This truly admirable little volume is made up of scattered fragments of instruction, furnished by the author in his capacity of tutor, to a pupil. It comprises a series of brief essays on such topics as these: On the Study of Language; Where had I best Travel; On Style; On English Composition; On Novels; How to give money away; A little Learning is not a dangerous thing, etc. They are written in a lively, easy style, and abound with practical suggestions and information of the highest value. The writer's power of illustrating and enforcing his precepts by the adduction of pertinent facts, is remarkable, and such as tutors are rarely blessed with. The essays on Style and English Composition are particularly worthy of perusal; and every page of the book contains matter that will profit not only the young but the old.—*Boston Commercial Bulletin.*

GLOBE EDITION OF SCOTT'S POETICAL WORKS.

The Poetical Works of Sir Walter Scott, Baronet, with a Biographical and Critical Memoir by Francis Turner Palgrave, late Fellow of Exeter College, Oxford. Square 12mo. Price, Cloth, $2.00; Cloth, extra gilt top, $2.25; Half Turkey, gilt top, $3.50; Half Calf, gilt extra, $4.00; Full Roxburgh Turkey, gilt extra, $6.00.

POEMS.

By Mrs. Frances Dana Gage. Printed on fine tinted paper. 12mo. Cloth, beveled boards. Price $1.75.

ELSIE MAGOON; or, The Old Still-House.

A Temperance Tale. Founded upon the actual experience of everyday life. By Mrs. Frances D. Gage. One vol. 12mo. Cloth. Price $1.50.

LAST DAYS OF A KING.

An Historical Romance. Translated from the German of Moritz Hartmann by Mary E. Niles. 12mo. Cloth. Price $1.50.

ROBERT SEVERNE: His Friends and his Enemies.

A Novel. By William A. Hammond. 12mo. Extra Cloth. Price $1.75.

www.ingramcontent.com/pod-product-compliance
Lightning Source LLC
Chambersburg PA
CBHW021505210326
41599CB00012B/1143